Tonsteins: Altered Volcanic-Ash Layers in Coal-Bearing Sequences

Bruce F. Bohor
U.S. Geological Survey
Branch of Coal Geology
MS 972, Box 25046, Federal Center
Denver, Colorado 80225

and

Don M. Triplehorn
Department of Geology and Geophysics
University of Alaska
Fairbanks, Alaska 99775

SPECIAL PAPER

285

1993

Copyright © 1993, The Geological Society of America, Inc. (GSA). All rights reserved. GSA grants permission to individual scientists to make unlimited photocopies of one or more items from this volume for noncommercial purposes advancing science or education, including classroom use. Permission is granted to individuals to make photocopies of any item in this volume for other noncommercial, nonprofit purposes provided that the appropriate fee ($0.25 per page) is paid directly to the Copyright Clearance Center, 27 Congress Street, Salem, Massachusetts 01970, phone (508) 744-3350 (include title and ISBN when paying). Written permission is required from GSA for all other forms of capture or reproduction of any item in the volume including, but not limited to, all types of electronic or digital scanning or other digital or manual transformation of articles or any portion thereof, such as abstracts, into computer-readable and/or transmittable form for personal or corporate use, either noncommercial or commercial, for-profit or otherwise. Send permission requests to GSA Copyrights.

Copyright is not claimed on any material prepared wholly by government employees within the scope of their employment.

Published by The Geological Society of America, Inc.
3300 Penrose Place, P.O. Box 9140, Boulder, Colorado 80301

Printed in U.S.A.

GSA Books Science Editor Richard A. Hoppin

Library of Congress Cataloging-in-Publication Data

Bohor, Bruce Forbes, 1932–
  Tonsteins : altered volcanic ash layers in coal-bearing sequences / Bruce F. Bohor and Don M. Triplehorn.
      p.   cm. — (Special paper ; 285)
  Includes bibliographical references.
  ISBN 0-8137-2285-3
  1. Tonsteins. I. Triplehorn, D. M. II. Title. III. Series: Special papers (Geological Society of America) ; 285.
  QE471.15.T64B64   1993
  552'.5—dc20   93-17797
                                                   CIP

Front cover: Four tonsteins (white weathering layers) in the C coal bed of the Ferron Sandstone Member, Mancos Shale (Upper Cretaceous) exposed along the Coal Cliffs in central Utah. Overlying massive marine sandstone is the basal unit of the subsequent deltaic cycle (see Ryer et al., 1980). Photograph by Tom Ryer. **Back cover:** Bohor on outcrop of the "Big Dirty" coal bed, which contains multiple crystal tonsteins (Fort Union Formation, Paleocene, near Roundup, Montana).

10 9 8 7 6 5 4 3 2 1

Contents

Abstract 1

Introduction 2

Nomenclature and Classification 3

Geographical Occurrence of Tonsteins 4

Volcanic Characteristics of Tonsteins 5

Methods of Study 5
    Disaggregation 5
        Dispersion 5
        Ultrasonic Treatment 6
        Mechanical Disaggregation 6
        Chemical Disaggregation 6
    Digestion in Hydrofluoric Acid 6
    Size Separation and Panning 7
    Polished Specimens 7
    Fractured Surfaces 7

Mineralogy 7
    Primary Minerals 7
        Quartz 7
        Feldspars 8
        Biotite 9
        Apatite 11
        Ilmenite and Magnetite 12
        Zircon 13
        Rutile and Sphene (Titanite) 14
        Other Primary Components 15
        Glass 15
    Secondary Minerals 15
        Kaolinite 15
        Smectite 16

Illite	16
Phosphates	17
Pyrite	18
Diatoms	18
Weathering Products	18

Gross Texture of Tonsteins 18
    Vermicules, Plates, and Microspherules 23
    Accretionary Lapilli 23
    Graupen 23
    Breccia 25
    Flint Clay 25
    Clay-Free Partings 25
    Glass Shards and Lapilli 26
    Rooting and Burrowing 26
    Stumps, Twigs, and Other Plant Materials 26
    Reworking 27

Alteration of Volcanic Ash to Tonstein 27
    Time of Alteration 27
    Influence of Bed Thickness 29
    Ash Composition 30
    Alteration Sequence 30

Environmental Effects of Thick Ash Falls 32
    Plants, Hydrology, and Coal Chemistry 33
    Secondary Minerals 33

Stratigraphic Uses of Tonsteins 33
    Radiometric Dating of Primary Minerals 34
    Calibration of Fossil Zones 34
    Isochron Horizons 34
    Correlation of Individual Units 36
        Lithologic Succession 36
        Field Appearance (e.g., Thickness, Color, and Texture) 36
        Bulk Clay Mineralogy 36
        Bulk and Trace-Element Chemistry 36
        Relative Abundance of Nonclay Minerals 36
        Elemental Composition of Mineral Species 36
        Morphology of Mineral Species 37
    Recording Eruptive Histories 37
    Intercontinental Correlations 38

Summary and Conclusions 38

Acknowledgments 40

References Cited 40

Tonsteins: Altered Volcanic-Ash Layers in Coal-Bearing Sequences

ABSTRACT

Volcanic ash that falls into marine settings commonly alters to smectitic deposits known as bentonites, the volcanic origin of which has been recognized for many decades. However, volcanic ash falling into nonmarine coal-forming environments generally alters to kaolinitic claystones called tonsteins, and these beds have only recently been universally accepted as being volcanic in origin. The recognition of tonsteins as altered volcanic ash is based on mineralogy, texture, radiometric age, and field relations.

Tonsteins occur on almost every continent, but are best known from Europe and North America. Their geologic range is coincident with that of coal-forming environments; i.e., from Devonian to Holocene. The coal-forming environment is well suited for preservation of thin air-fall deposits because it features low depositional energy, topographic depression, rapidity of burial by organic matter, and lack of detrital input due to the baffling effect of plant growth. Volcanic ashes deposited within or beneath peat beds are strongly affected by humic and fulvic acids generated from organic matter. This acidic, organic-rich, highly leaching environment is partly responsible for the alteration of volcanic glass and mineral phases into kaolinite by first-order (solution-precipitation) reactions. Bed thickness also affects ash alteration, resulting in a vertical zonation of clay mineralogy in thick beds. In addition, voluminous ash falls can have an important effect on the biological and hydrological regimes of the peat swamp.

Most distal tonsteins contain a restricted suite of primary volcanic minerals, such as euhedral beta-quartz paramorphs and water-clear quartz splinters (both with glass inclusions), sanidine, idiomorphic zircon, biotite, rutile, ilmenite and magnetite, apatite, allanite, and other accessory minerals specific to a silicic magma source. Textural features indicating an volcanic air-fall origin include bimodal size distribution of components, "graupen," accretionary lapilli, altered glass bubble junctions, and aerodynamically shaped altered glass lapilli. Radiometric dating of primary minerals in tonsteins shows that they are coeval with the stratigraphic ages of enclosing rocks. Tonstein field relations indicate an volcanic air-fall origin because they are thin, widespread, continuous layers, with sharply bounded upper and lower contacts, that often pass beyond the bounds of the swamp and are occasionally penetrated by stumps in growth position.

The volcanic air-fall origin of tonsteins predicates their usefulness in many geologic studies. Because they are isochronous, tonsteins can be used to vertically zone coal beds and thus provide controls for geochemical sampling, organic petrography studies, and mine planning. Regional correlations of nonmarine strata can be made with tonsteins, and intercontinental correlations may be possible. Furthermore, the presence of clay-free volcanic-ash layers in coal beds may indicate a raised-bog ori-

Bohor, B. F., and Triplehorn, D. M., 1993, Tonsteins: Altered Volcanic-Ash Layers in Coal-Bearing Sequences: Boulder, Colorado, Geological Society of America Special Paper 285.

gin for the peat swamp. Radiometric dating of primary volcanic minerals in tonsteins allows age determination of coal beds and the calibration of palynomorphic zones. Multiple tonsteins in thick coal beds may be useful for studying the style and history of explosive volcanism.

INTRODUCTION

Recent distal volcanic ash fallout deposits are well-preserved in marine environments and have been studied extensively (cf. Kennett, 1981). Ancient analogs of these deposits are preserved as altered ash beds called bentonites, initially described from the Benton Group in Montana (Ross and Shannon, 1926). The glassy phase of these marine ash beds commonly alters to the smectite clay minerals, including montmorillonite. Distal ash fallout deposits also may be preserved on land surfaces. Recent eruptive events are sometimes preserved as unaltered ash layers that often can be correlated over great distances (Izett, 1981). Ancient nonmarine ash layers are not as well preserved, however, because of erosional processes acting on subaerial surfaces prior to and following burial. The depositional environments best suited to preservation of volcanic ash fallout deposits on land are shallow lakes, ponds, and swamps. In these environments, low depositional energy, plants acting as clastic sediment baffles, subdued topography, and rapid burial by organic materials (peat) favor the preservation of ash in a relatively pristine state. Therefore, volcanic ash falling on land surfaces is commonly preserved in these environments, along with the peat normally developed and preserved here. With time, compaction and diagenesis act on these preserved volcanic ash and peat layers to convert them into beds of tonstein and coal, respectively.

Tonsteins were accepted as altered volcanic ash only relatively recently, and then only after much argument and dissension. One of the main reasons for this dissension was the commonly held assumption that volcanic glass alters exclusively to smectite (the bentonite scenario), whereas the clay partings preserved in coal-forming environments are mainly kaolinitic. Clay mineralogists, however, have determined that volcanic glass alters to whatever clay mineral is stable in the depositional environment at the time of alteration (Keller, 1970). As Schultz (1963) pointed out, even volcanic ash deposited in marine environments can alter to kaolinite occasionally, so that bentonites are not necessarily always smectitic. In acidic peat swamps, the formation of kaolinite usually is strongly favored by these geochemical conditions and thus volcanic ash is altered and preserved here primarily as kaolinitic tonsteins, although smectitic tonsteins are found occasionally.

The study of tonsteins began in Germany, particularly in the Upper Carboniferous coal beds of the Saar region. A volcanic origin for these layers was proposed near the turn of the century by Schmitz-Dumont (1894). Many other European workers subscribed to the volcanic theory in the following half-century, culminating in an excellent monograph detailing the volcanic nature of tonsteins published on the occasion of the First International Colloquium on coal tonsteins held in Ostrava, Czechoslovakia (Dopita and Kralik, 1977). While evidence grew for the volcanic theory, other workers, lead by Karl Hoehne in Germany, presented a dissenting view of tonsteins as products of weathering of ordinary sediments in a swamp environment (Hoehne, 1953a). Hoehne published voluminously on tonsteins from 1944 until his death in 1962. Hoehne's eloquence and productivity managed to draw many converts to the sedimentary theory, but eventually the pendulum of scientific opinion swung back to the volcanic scenario. An important contribution to the resurgence of the volcanic theory was the discovery of younger, less-altered tonsteins containing unequivocal primary volcanic minerals and even volcanic glass (Triplehorn and Bohor, 1986). Hoehne presented a synopsis of the development of the volcanic hypothesis and its major adherents and defended his sedimentary theory in his final paper, published posthumously (Hoehne, 1964). The volcanic theory of the origin of tonsteins is now accepted almost universally, even by the most dedicated of Hoehne's disciples (cf. Burger, 1985b).

The confirmation of tonsteins as nonmarine altered volcanic-ash layers analogous to marine bentonites was extremely important for establishing their geologic uses. Masek (1963) proved the genetic relation between these two types of ash layers when he traced the lateral transformation from bentonites to coal tonsteins in the central Bohemian basin. Waage (1961) also traced kaolinitic noncoal tonsteins in deltaic facies of the Lower Cretaceous Dakota Group into marine smectitic bentonites along the northern Front Range foothills in Colorado. This equivalence between these two types of volcanic layers indicates that tonsteins can be used in the same ways that bentonites have long been utilized, i.e., as isochrons, marker horizons, calibrators of fossil biozones, and sources of primary volcanic minerals suitable for radiometric dating.

We are assuming here that the volcanic theory of tonstein origin is now accepted as fact, because all of our data, as well as those of most other workers, strongly support this theory. On this basis, we present summaries of the results of our studies on tonsteins over the past 15 years, with added references to others who have made contributions to the subjects under discussion. Thus, the ideas presented about tonsteins are mostly our own, and not just a consensus review of the field. Some of our data and procedures have not been published previously, and are being presented here for the first time. We have attempted to gather together into one place everything pertaining to the characterization and usage of tonsteins, so that these partings can assume their rightful place as a powerful tool for geologic studies.

NOMENCLATURE AND CLASSIFICATION

Herein, the term "tonstein" refers to nonmarine, generally kaolinitic layers derived from the in situ alteration of air-fall volcanic ash. The genesis of the name tonstein was discussed by Williamson (1970a) and by Bouroz et al. (1983). It is a German term (literally, "claystone") that originally had no genetic connotation (Bischof, 1863), but we restrict it here to altered volcanic-ash layers found in nonmarine environments commonly associated with coal formation, usually kaolinitic in composition. Kaolinitic volcanic-ash layers occasionally found in marine environments (Schultz, 1963; Pollastro and Martinez, 1985) are not tonsteins by our classification, but smectitic ash layers occurring in nonmarine, generally coal-bearing environments are. Other names have been proposed for ancient nonmarine volcanic-ash layers, such as cinerites (Bouroz, 1962) and kaolinitic bentonites (Spears and Rice, 1973), but tonstein seems to be more accepted and widely used as a general field term. Laboratory analyses usually are required to confirm the volcanic origin of a tonstein, but almost all of the kaolinitic partings in coal beds of the United States have proven to be volcanic. Therefore, linking a volcanic ash-fall origin to the field term tonstein is not unreasonable.

Many classification schemes have been applied to tonsteins (see discussions in Stach et al., 1982; Bouroz et al., 1983; Burger, 1985a), but most of these schemes are descriptive and based generally on texture, clay-mineral grain morphology, and pseudomorphs; i.e., features mainly related to postdepositional processes and not to origin. Because these features are secondary and often vary laterally within a single layer, they contribute little to the genetic or mineralogic understanding of tonsteins. Our experience suggests that a general dichotomy into the categories of coal and noncoal tonsteins, with descriptive modifiers based on major components and grain size, such as crystal, graupen, vitric, coarse or fine grained, and microcrystalline, is adequate for most field purposes. Previous classifications based descriptions solely on clay textures, terms such "crystal" referring to large crystals or vermicules of kaolinite. In contrast, we use the term crystal to describe a tonstein whose ratio of primary volcanic phenocrysts (crystals) to clay is very high (Fig. 1; Plate IA). In like fashion, the term vitric is applied to a fine-grained tonstein whose ratio of volcanic phenocryst grains to clay is very low; the clay component is presumed to represent an altered glassy (vitric) phase. The field designation coal tonstein means that the tonstein layer is wholly contained within a coal bed, or is in direct contact with it. Burger (1979) used the term "kaolin-coal tonstein" for these layers, but this designation would exclude those tonsteins in coal beds that are not primarily composed of kaolinite. Noncoal tonstein refers to those nonmarine altered volcanic-ash layers that are not within or in contact with coal beds. A coal tonstein can pass into a noncoal tonstein when the former extends laterally beyond the coal swamp, or when it climbs above or drops below a coal bed into adjacent strata. That tonsteins have been observed in these noncoal depositional settings proves that they have no direct genetic connection to a peat swamp, such as a precipitate from swamp water, as was earlier suggested by Hoehne (1953a). Burger (1955, 1958) traced coal tonsteins in the Ruhr into the country rock beyond the bounds of the coal bed, confirming their independence of the coal-forming swamp regime.

Figure 1. Crystal tonstein. Thin-section photomicrograph (plane polarized light) showing high proportion of primary volcanic crystals (white) of quartz and feldspar, with very little clay (altered glass) in matrix; similar terminology is used to describe crystal tuff. Fort Union Formation (Paleocene), Roundup, Montana. Field width is 3.55 mm; also see Plate IA).

The term "cinerite," proposed by Bouroz (1962), applies generally to any air-fall volcanic-ash deposit regardless of its state of alteration, depositional environment, or present mineralogic composition. Thus, unaltered volcanic ashes, marine smectitic bentonites, and nonmarine kaolinitic tonsteins all would come under the general heading of cinerites. In this hierarchy, a volcanic-ash layer in a coal bed that has altered to kaolinite would be referred to as a cinerite first, a tonstein second, and a kaolin-coal tonstein as the last, most definitive descriptive term. In like manner, an altered smectitic volcanic ash in a marine depositional environment would be designated primarily as a cinerite, and secondarily as a bentonite; unaltered and partially altered marine or nonmarine volcanic ashes still could be classified as cinerites. The use of cinerite as the primary term in this hierarchy has the advantage of being the most general and all encompassing, while still retaining the important genetic connotation of an air-fall volcanic origin—a feature of great significance to geoscientists.

Cinerite is almost too broad a classification to be of much use, however, and has not been widely accepted. Tuff, a general term for consolidated volcanic ash, is already in common use (Fisher and Schmincke, 1984), and probably should be retained for those nonmarine ash layers not closely associated with coal. Once the depositional environment is identified, an ancient ash layer (cinerite) can be readily classified in the field as altered tuff, tonstein, or bentonite. Compositional modifiers based on clay mineralogy and textural components can be added later, after microscopic examination and lab analyses provide these data.

Fisher and Schmincke (1984) suggested that the term tonstein be eliminated and that the term bentonite be used in a broad sense to represent all thin, widespread, clay-rich layers of probable volcanic origin, regardless of their clay-mineral composition or depositional environment. This usage raises obvious problems, one of which is that the entrenched concept of bentonite in geology and industry as a primarily smectitic marine rock is at historical variance with it. Chamley (1989) concurred with this proposal by Fisher and Schmincke, but we disagree with this use of the term bentonite because of precedence and applicability. We suggest that the genetic term cinerite might be better suited to the top of the nomenclature hierarchy, with the specifically descriptive term bentonite restricted solely to marine altered volcanic ashes usually, but not invariably or exclusively, of smectitic composition.

Some preserved volcanic-ash beds seem to defy simple classification by our scheme. An example is a partially altered ash bed in the Troublesome Formation (Oligocene and Miocene) of Colorado (Zielinski, 1982). The Troublesome Formation was deposited in an alluvial and lacustrine environment that was not peat accumulating (at least none is preserved as coal or lignite). The basal portion of this ash bed is altered to smectite (montmorillonite), but the upper portion consists of unaltered glass and primary volcanic minerals. This bed is obviously a cinerite; further attempts at classification might designate it as a partially altered smectitic tuff. Had this ash layer been deposited in a peat swamp and entirely altered to smectite or kaolinite, it would be classified as a smectitic or kaolinitic tonstein. Examples of smectitic and kaolinitic tonsteins in various stages of alteration have been found in Miocene coal beds on the Kenai Peninsula, Alaska (Triplehorn et al., 1977), and partially altered marine ash beds in North Dakota as old as latest Cretaceous have been reported by Forsman (1984).

Degradation of organic matter in a peat swamp forms humic and fulvic acids, which promote alteration of volcanic ash to kaolinite (La Iglesia and Van Oosterwyck-Gastuche, 1978). However, organic matter (e.g., algae) in shallow-marine environments can also promote kaolinization of volcanic ash, as shown by the following examples. Kaolinitic beds, altered in situ from volcanic ash in Ordovician strata (preceding the evolution of land plants) interpreted as littoral and shallow marine, were called tonsteins by Garcia-Ramos et al. (1984); we would classify these beds as kaolinitic bentonites. Other examples of volcanic ashes (Cretaceous) altered to kaolinite in organic-rich marine environments were reported by Schultz (1963), Pollastro (1981), and Pollastro and Martinez (1985). These altered marine ashes are also designated as kaolinitic bentonites in these papers, conforming to our classification scheme.

GEOGRAPHICAL OCCURRENCE OF TONSTEINS

Tonsteins occur on all continents that contain coal beds. They were originally described from the coal beds of west-central Europe (Stutzer, 1931; Hoehne, 1949; Stach, 1950; Scheere, 1956; Bouroz et al., 1953), and later reported from Russia (Zaritsky, 1967; Tereshenko and Chernovyantz, 1979) and England (Williamson, 1961; Eden et al., 1963). Hoehne described tonsteins from northern Mexico (Hoehne, 1953a), India (Hoehne, 1953b), Australia (Hoehne, 1957), and western Canada (Hoehne, 1959). Other worldwide occurrences reported by various workers include Indonesia (Addison et al., 1983), China (Zhou et al., 1982), Japan (Bouroz, 1962), South Africa (Spears et al., 1988), Spain (Prado, 1964), Colombia, South America (Lambrecht and Scheere, 1965), Australia (Diessel, 1963; Loughnan, 1971), Antarctica (J. Thorez, 1991, personal commun.), and western Canada (Meriaux, 1972; Spears and Duff, 1984). Tonsteins have also been reported from Brazil, Bulgaria, Argentina, Rhodesia, and Turkey (Burger, 1979). This listing is not meant to be exhaustive, but illustrates generally the broad distribution of tonstein occurrences throughout the world. Burger (1985a) presented an extensive listing of literature sources on tonsteins in western European coalfields, and Burger and Damberger (1985) listed those known from North America at that time.

In the United States, the first report of a tonstein (although not designated as such) was made by Rogers (1914) from a coal bed in the Lance Formation, Montana. Rogers recognized the probable volcanic origin of this kaolinitic parting, a remarkably prescient concept. Seiders (1965) attributed a volcanic origin to a flint clay (actually a recrystallized tonstein) from a Pennsylvanian coal bed in Kentucky. Stevens (1979), and later Bohor and Triplehorn (1981), confirmed the volcanic origin of this parting and classified it as a tonstein. Bohor and others (1976) and Bohor and Pillmore (1976) reported tonsteins in Mesozoic and Cenozoic coal beds of the Rocky Mountain region. At about the same time, Triplehorn (1976a, 1976b) described altered ash beds from Alaskan coals and began dating them, as well as similar beds discovered in western Washington, by radiometric measurements of their primary volcanogenic minerals (Triplehorn et al., 1977, 1984). Triplehorn and Bohor (1981) described tonsteins in coal beds of the Ferron Sandstone Member of the Mancos Shale in central Utah, and Ryer et al. (1980) showed how a thick coal bed in this unit could be zoned using tonsteins as isochronous marker horizons. Senkayi et al. (1984, 1987) described tonsteins from Eocene coal beds of the Texas Gulf Coast region. Triplehorn et al. (1989) reported several tonstein horizons in

Middle Pennsylvanian coal beds of the Appalachian basin. Tonsteins also occur in the Illinois and Forest City basins of the mid-continent region in Carboniferous strata, and in Cretaceous strata of the Kaiparowits and Black Mesa coal fields of southwestern United States (Bohor, unpublished data).

VOLCANIC CHARACTERISTICS OF TONSTEINS

One characteristic of tonsteins that suggests an air-fall origin is their field relations. Tonstein layers are usually thin, continuous, widespread, and have sharp contacts with the enclosing beds. All these features characterize a volcanic fallout deposit preserved in a low-energy environment. The lateral extent of these layers is dependent upon that of their preservational environment, as well as the areal dimensions of the volcanic plume from which they were derived. Most coal tonsteins weather white on outcrop because darker organic materials have been bleached out, exposing the natural white color of the kaolinite (see front cover photo). However, in cores or on fresh surfaces in mines, coal tonsteins are often brown, gray, or black.

The identification, definition, and geologic uses of tonsteins are all based on a volcanic-ash origin. Therefore, it is useful to list the characteristics of tonsteins that serve to define their origin as altered volcanic ash. Most of the characteristics listed here are those suggested by Triplehorn and Bohor (1981), but other workers had identified some of them earlier for bentonites (Ross, 1928) as well as for tonsteins.

1. Thin, widespread, relatively uniform bed character, with sharp bounding contacts and showing no evidence of water transport.
2. Marked bimodal size distribution; i.e., sand-sized volcanic mineral grains (phenocrysts) floating in a clay-sized matrix.
3. A limited mineralogical suite of euhedral or broken primary volcanic phenocrysts, including (1) water-clear, beta-form quartz-crystal paramorphs and quartz splinters, both often bearing glass inclusions; (2) high-temperature, K-feldspar (sanidine) crystals in rhyolitic ashes (Ca-plagioclase often occurs in dacitic ashes); microcline is *not* a primary volcanic mineral; (3) idiomorphic zircon crystals with a limited range of morphologies from individual beds; (4) magnetite-ilmenite, biotite, and occasionally apatite; and (5) trace volcanic minerals, such as allanite and sphene.
4. Nearly monmineralic clay-mineral composition, usually kaolinite (detrital suites are almost always heterogeneous), with sharp X-ray diffraction peaks (authigenic).
5. Delicate kaolinite vermicules indicating in situ alteration, not transportation of detrital grains by water.
6. Radiometric ages of primary minerals are coeval with biostratigraphic ages of the enclosing sediments, which would not be the case if they were detrital.
7. Preservation of accretionary lapilli and pumice (graupen?) as relict structures.
8. Preserved volcanic glass textures, such as shards and bubble junctions.
9. Lesser quartz content than that of normal shales.
10. Extension of tonstein layer beyond the swamp environment.
11. REE chondrite-normalized patterns similar to those of rhyolitic ashes, including large negative europium anomalies.

Individual characteristics from this list are not necessarily specific to tonsteins, but if they apply as a group, then they serve to define the volcanic origin of a parting and allow its classification as a tonstein. We discuss aspects of these characteristics in detail in subsequent sections.

METHODS OF STUDY

Conventional techniques applied to tonsteins in the past have included bulk X-ray diffraction (XRD), bulk chemistry, macroscopic description, and thin-section petrography. In this section we describe some additional techniques that have proven particularly useful in characterizing tonsteins. For the most part, these techniques have not been described previously in the literature.

Our approach to the study of tonsteins differs from that of other workers by its emphasis on the coarse-grained (clay-free fraction. We concentrate on the mineralogy, composition, and morphology of individual mineral grains in this fraction. We developed techniques to separate clay from sand- and silt-sized material in order to characterize specifically the primary volcanic mineral component (Triplehorn and Bohor, 1981). Extensive use was made of the scanning electron microscope (SEM) and grains mounted in immersion oils for this characterization. The identification of primary minerals can contribute to the distinction of individual tonsteins, determination of their volcanic origin, interpretation of eruptive style and magma composition, and radiometric dating. Authigenic and secondary minerals, such as pyrite, alumino-phosphate minerals, and kaolinite vermicules, may be important for interpreting diagenesis; organic components, such as large spores and resin rodlets, are commonly found in the coarse fraction.

Disaggregation

A major objective of our tonstein studies is to obtain a clay-free coarse fraction with minimal effort, but without destroying any primary mineral phenocrysts in the process. This objective is relatively easily attained with younger, less-altered tonsteins and those of smectitic compositions, but disaggregation of hard, fine-grained, flint-like kaolinitic tonsteins can be very difficult. The following procedures are arranged roughly in order of increasing effort and deleterious effects on mineral components.

Dispersion. The crushed sample is soaked in water and agitated to suspend the clay fraction; the suspension is decanted off after a suitable period of settling determined by

Stoke's Law. Mechanical stirring in a blender can be employed when hand stirring is no longer effective. The addition of dispersing agents, such as Calgon or sodium hexametaphosphate, may help disperse and suspend the clay fraction. Standard mud-disaggregation techniques similar to these were described by Folk (1980).

Ultrasonic treatment. Ultrasound dispersion can be used in combination with the procedures described above. This treatment normally produces no adverse effects on primary minerals, but prolonged and intense treatment can damage fragile structures and round off soft components. These treatments are usually adequate to disperse the clay for removal by elutriation, except in the case of tonsteins that contain large single crystals of kaolinite, vermicules, or flint-like aggregates of fine-grained kaolinite.

Mechanical disaggregation. Some kaolinitic aggregates resist complete breakdown in the blender and ultrasonic bath. When these methods are no longer effective in suspending clay and breaking down aggregates of kaolinite, the residue may be placed in a mortar with a small amount of water covering the sample and rubbed gently with a rubber racquetball or squash ball. This technique breaks up the clay aggregates without crushing or abrading quartz grains and other phenocrysts.

Chemical disaggregation. Some kaolinitic tonsteins are not susceptible to complete disaggregation by the mechanical means described previously. Very fine grained to microcrystalline tonsteins are composed of interlocking kaolinite crystals, and samples with this texture will not disaggregate without more intense mechanical shearing. This texture is also found in flint clays that are characterized by smooth, dense, flint-like surfaces and conchoidal fracture. Some older tonsteins that have undergone recrystallization display this texture and can be classified as flint clays (Bohor and Triplehorn, 1981).

Other methods of disaggregation are required for these difficult samples that resist ordinary techniques. The most effective method is chemical disaggregation, where a lightly crushed sample of flint-like clay is soaked in organic intercalating agents, such as dimethyl sulfoxide (DMSO) or hydrazine monohydrate. These compounds penetrate between the kaolinite platelets, thereby weakening intercrystallite bonds and promoting disaggregation through swelling pressures. The pertinent literature (Weiss et al., 1963; Olejnik et al., 1968) discussed only the preparation of clay-organic intercalations with these solvents, and did not address disruptive effects on clay aggregates. DMSO and hydrazine monohydrate are very effective in disaggregating kaolinitic clays that are resistant to the purely mechanical treatments described previously.

No detailed controlled experiments have been reported on the best applications of organic compounds to kaolinite disaggregation. However, such experiments have been carried out (Triplehorn and others, 1993), and the following scheme is a condensed procedure based on these results.

If no mechanical disaggregation is desired, soak the coarsely crushed, water-wetted sample of resistant clay in an excess of DMSO in a sealed container for about three to four weeks; similar results can be expected from hydrazine monohydrate applied to crushed and dried clay in about one to three days. Olejnik et al. (1968) found that dilution with 8% water enhanced DMSO intercalation of kaolinite. Sequential treatment with DMSO and hydrazine monohydrate may also be more effective than either of these solvents alone (Calvert, 1984). These techniques reduce the clay to a paste-like consistency that can be readily dispersed in water, allowing decantation of the suspended clay fraction. Some fine granular kaolinite aggregates and tiny vermicules resist this treatment and remain in the residue; these can be removed by hydrofluoric acid (HF) treatment if their presence would inhibit study of the primary mineral grains.

Some precautions should be taken when using these liquids. Hydrazine monohydrate is relatively stable, but it can be explosive under certain conditions; it is also highly alkaline and caustic. DMSO is stable, but somewhat flammable; it readily penetrates the skin, which could introduce harmful substances into the bloodstream. The manufacturer's chemical-hazard warning literature should be obtained and read thoroughly before using these substances.

Digestion in hydrofluoric acid

Treatment with HF is effective in removing kaolinite from the coarse-grain residue, but may also attack quartz and other silicates. Therefore, it should only be used when a completely clay-free mineral residue is sought, or when an immediate separation of resistates is required. Acid concentration and treatment time are critical and vary with the sample. Euhedral beta-quartz crystals and thin sharp splinters of quartz showing no apparent sign of solution have been recovered after an hour in dilute HF when an excess of kaolinite was present during the treatment. The HF should be neutralized by dilution with water as soon as the last remnants of clay are dissolved. Any carbonate present should be removed with dilute HCl prior to HF treatment to avoid precipitating insoluble fluorides (see Norrish, 1968, for further discussion).

It is sometimes useful to etch large tonstein fragments or polished surfaces with HF, causing less soluble mineral grains to stand out in relief from the clay matrix. This procedure clearly reveals the distribution and orientation of the primary volcanic grains with respect to the altered glass matrix, as well as their morphology. The most effective treatment parameters for different samples are best determined by trial and error. We recommend starting with a solution of 50% HF (by volume) for one hour; examine the result and adjust accordingly. Avoid applying concentrated HF to dry, fine-grained clay, because this may result in a dangerously rapid reaction (boiling and splattering). (Always use HF in a fume hood and wear adequate skin and eye protection—HF is extremely hazardous!)

Size separation and panning

Once all the clay has been removed, the sand-sized fraction is ready for visual examination and separation of individual grains for microphotography, SEM, XRD, chemistry, or age dating. Because mineralogy varies with grain size, the dried coarse residue should be passed through a series of 3 in. sieves (425, 250, 125, and 75 μm openings). Each size fraction can then be examined independently in standard flat-bottomed petri dishes.

Crude heavy-mineral separations of mineral residues are possible by panning individual size fractions in a petri dish. Coal and other low-density material can be decanted in the process. Swirling or oscillating motions usually concentrate grains of different densities and shapes in different areas of the dish. Some experimentation and practice may be necessary to achieve good results. Careful removal of the water with a syringe, followed by rapid drying in a microwave oven or under a heat lamp, permits hand picking of dried grains from the concentrate. If larger quantities of heavy minerals are needed, this procedure can act as a screening tool, followed by conventional heavy-liquid separation techniques.

Magnetic minerals are detectable by passing a small magnet under the petri dish on a binocular microscope stage. The grains that show movement in response to the magnet can be recovered with a magnetized needle or probe. Covering this device with a sleeve of plastic or aluminum foil allows the complete release of magnetic grains from the probe.

Polished specimens

Hard, dense, kaolinitic tonsteins can be polished by conventional methods. Impregnation with acetate or imbedding in casting resins may facilitate handling and improve the durability of fragile samples. For small samples, procedures for the preparation of coal samples outlined by Stach et al. (1982) are recommended.

Polished specimens are examined using reflected light, either with a binocular microscope or using reflected-light microscopes routinely employed in ore microscopy or coal petrology. Etching with HF may also be useful for enhancing relief and textural contrast. Ultraviolet light, commonly used in coal petrography, may reveal fluorescent minerals or organic constituents. Polished surfaces are also required for examination with the electron microprobe.

Fractured surfaces

Naturally fractured or etched surfaces can be examined with a scanning electron microscope. Element mapping (e.g., Spears et al., 1988) is useful for identifying small components or studying the distribution of individual elements. The backscatter mode of the SEM provides images with variable contrast levels related to atomic number. Using this technique, Triplehorn and Finkelman (1989) were able to photograph textural components, particularly alumino-phosphate psuedomorphs of glass shards, only marginally visible optically in thin section or on polished surfaces.

MINERALOGY

The distal volcanic ash falls forming most tonsteins are derived mainly from rhyolitic magmas, because only the explosive eruptive style of silicic volcanism provides the energy necessary to loft tephra to heights compatible with extended downwind transport. Some rhyolitic ash falls contain mostly glass; others include moderate to abundant amounts of magmatic minerals as individual grains (phenocrysts). Rarely, mineral grains and rock fragments plucked from country rock lining the throat of the volcano are found in tuffs. All such minerals grains are here termed "primary." In contrast, "secondary" minerals are those formed after deposition of the tuff, either by alteration of original ash components or by precipitation of new material from ground or surface waters during diagenesis or weathering.

The primary minerals characteristic of siliceous ash falls have been identified from modern tephras (cf., Izett, 1981). Most nonresistate primary minerals are either absent or present only in relatively low abundance in tonsteins, having been removed by weathering and diagenesis. We discuss only those that commonly survive these processes in the formation of tonsteins.

Primary minerals

Quartz. Generally, quartz is the most common and abundant primary mineral present in tonsteins. It crystallizes as phenocrysts from silicic magmas in the high-temperature (beta) form, inverting to the low-temperature phase (alpha paramorph) when temperatures fall below 573 °C. The occurrence of volcanic quartz in the euhedral bipyramidal form was described by Blatt et al. (1980). These authors pointed out that the quartz phase in extrusive and hypabyssal rocks crystallizes in a fluid, volatile-rich environment conducive to the development of crystal faces (euhedral morphology). In the more viscous, crystal-rich plutonic settings, however, the development of crystal faces is restricted by growth of neighboring grains, resulting in embayed, anhedral quartz forms typical of granites. Volcanic quartz is characteristically "water-clear," but often contains glass, fluid, or mineral inclusions (Clocchiatti, 1975).

Whole, euhedral, beta-quartz crystals are common in tonsteins, but more often only one crystal face occurs on otherwise sharply angular to cuspate grains; presumably these fragments represent broken bipyramids. Beta-form quartz crystals in tonsteins range from sharply euhedral bipyramids with no prism development (Fig. 2, A and B), or bipyramidal crystals slightly modified with minor prismatic faces (Fig. 3), to beta-quartz crystals rounded to various degrees, including almost spherical forms (Fig. 4, A and B). This rounding is due to resorption of

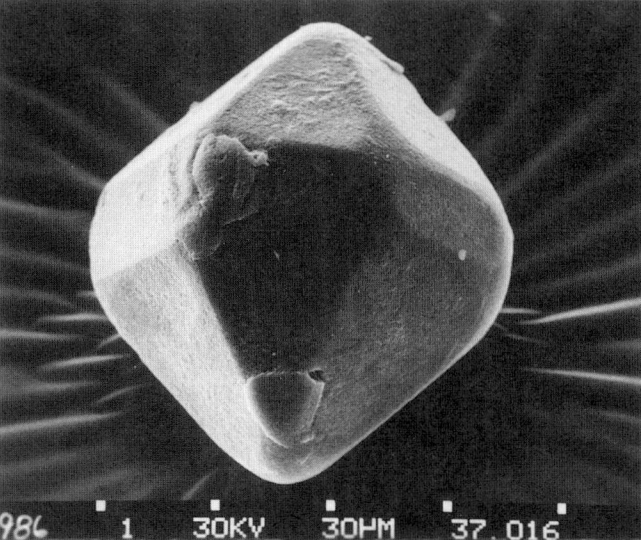

Figure 3. SEM photo of slightly rounded beta-form quartz crystal displaying minor prism face development. Crystal is from the Fire Clay parting, Hazard #4 coal bed (Middle Pennsylvanian), Breathitt Formation, eastern Kentucky.

Figure 2. Euhedral beta-form (high) quartz paramorph crystals having bipyramidal morphology and lacking prismatic faces, SEM photos. A: Three crystals from the Ruffner flint clay (Middle Pennsylvanian), near Charleston, West Virginia. B: Single crystal from the Fire Clay parting, Hazard #4 coal bed (Middle Pennsylvanian), Breathitt Formation, eastern Kentucky.

the edges of euhedral beta-quartz crystals within the magma chamber (Donaldson and Henderson, 1988), and should not be confused with detrital abrasion.

Heiken and Wohletz (1985) described quartz grains from a tuff as irregularly shaped, with delicate sharp edges and clean surfaces, and commonly conchoidally fractured. Thorez and Pirlet (1979), among others, described quartz in a tonstein as "shards" or "splinters." Splinter quartz seems to be the better term, avoiding confusion with the well-established usage of "shard" as applied to shattered glass components in tephra, and with "sherd" applied to archeological ceramic fragments. Such blade- and flake-shaped quartz splinters having thin, sharp edges (Fig. 5, A and B) are not generally characteristic of most fluvial or eolian terrigenous detritus.

The volcanic origin of euhedral beta-form quartz crystals in tonsteins has been questioned because of reported authigenic quartz grains in coal (Ruppert et al., 1984). However, in this occurrence the authigenic quartz is not euhedral and lacks the beta form. Further evidence of the volcanic origin of bipyramidal (beta) quartz crystals in tonsteins comes from analyses of glass inclusions within them (Fig. 6 and Plate IB) that show the composition of the parent magma to have been rhyolitic (Belkin and Rice, 1989). Cathodoluminesence of these crystals (Triplehorn and others, 1991) also confirms their volcanic origin.

Feldspars. The next most abundant primary minerals in tonsteins after quartz are feldspars. Plagioclase (Fig. 7) and the high-temperature variety of potassium feldspar, sanidine (Figs. 8 and 9), are common and sometimes relatively abundant; microcline is not a primary volcanic mineral. However, because feldspars are more susceptible to alteration than quartz, their abundance in tonsteins may not reflect their original abundance in the parent volcanic ash. Moreover, the relative amounts of different kinds of feldspar in tonsteins may be modified from those of the parent ash because plagioclase, particularly the calcium-rich variety, is apparently much more readily altered or dissolved during weathering than is sanidine (Banfield and Eggleton, 1990).

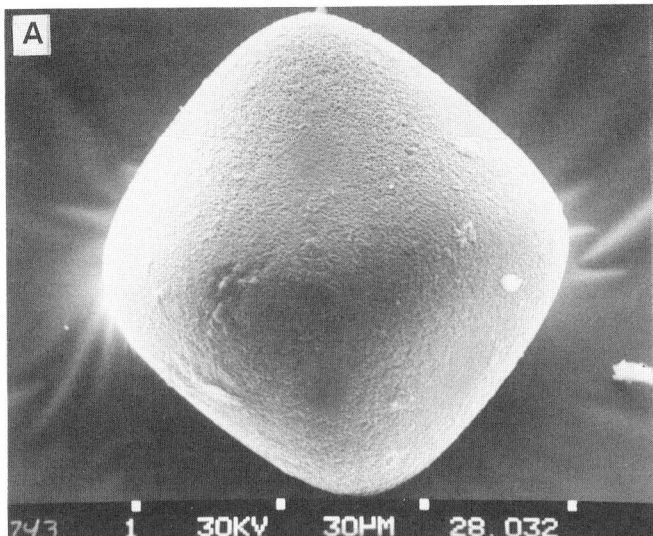

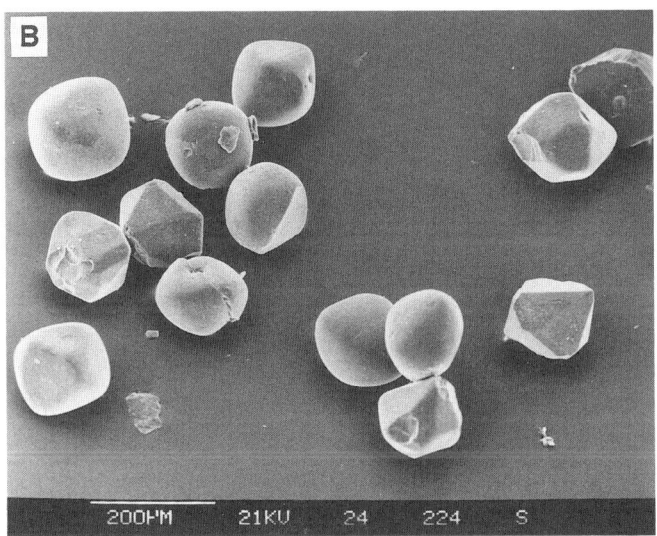

Figure 4. SEM photos of rounded beta-form quartz crystals. A: Rounded and euhedral beta-form quartz crystals from the upper part of the Denver Formation (Paleocene), Kiowa County, Colorado. B: Rounded and euhedral beta-form quartz crystals from an unnamed coal bed in the Dakota Formation (Upper Cretaceous), Henrieville area, Utah.

Figure 5. SEM photos of quartz splinters and flakes. A: Bald Knoll coal bed, Dakota Formation (Upper Cretaceous), Henrieville area, Utah. B: Quartz flake (left) and rounded beta–quartz (right) from a tonstein (Upper Permian) in Yunnan Province, China.

Fresh volcanic feldspar, when free of inclusions and argillic alteration, may strongly resemble quartz (Dopita and Kralik, 1977); thus, shape alone is usually not a reliable identifying characteristic. Quartz can best be distinguished rapidly from sanidine in the petrographic microscope by refractive index (R.I.) measurements, quartz having a higher R.I. than sanidine. Selection of an immersion oil with an R.I. between that of quartz and sanidine (i.e., 1.53) will separate these two types of grains by their color fringes. This technique facilitates separation of sanidine grains from clay-free residues of tonsteins for radiometric dating.

Biotite. This mica occurs as shiny black hexagonal flakes or stacks, often sharply euhedral (Fig. 9); it is a common component of silicic volcanic ashes. Biotite is less common in tonsteins than its abundance in fresh volcanic ashes might suggest, because it is one of the most readily altered of the primary volcanic minerals in the coal-forming environment. Muscovite is not a primary volcanic mineral, and its presence in tonsteins usually indicates detrital contamination.

Biotite commonly alters to kaolinite in tonsteins. Biotite flakes, bleached and altered to silvery or colorless kaolinite flakes without any apparent change of thickness or microstructure, strongly resemble muscovite upon visual examination.

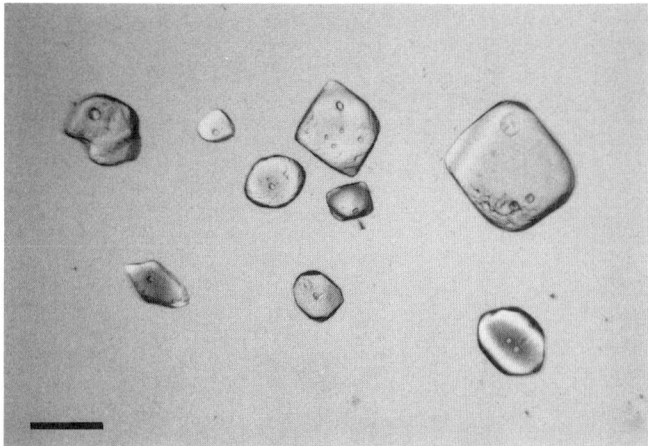

Figure 6. Beta-form quartz paramorph crystals (some rounded) and quartz fragments containing glass inclusions from bentonite at the Permian-Triassic boundary in China; scale bar = 200 μm. Compare with glass inclusions in beta-quartz crystals from tonsteins (see Plate IB).

Figure 7. SEM photo of plagioclase feldspar showing twin lamellae, from a smectitic tonstein in the Skookumchuck Formation (Eocene), Centralia, Washington. Note solution etching on edge of crystal.

Figure 8. SEM photos of euhedral sanidine crystals (A, B) from the thick tonstein in the C coal bed, Ferron Sandstone Member, Mowry Shale (Upper Cretaceous) in central Utah. Note monoclinic forms and lack of weathering.

Alteration of biotite stacks often proceeds from the edges inward, with bleaching of the dark color and leaching of iron, magnesium, and potassium from the rims (Fig. 10); the marked swelling of the rims results in doughnut-shaped grains (Fig. 11 and Plate IC). This type of biotite alteration was described by Bouroz et al. (1983), Dopita and Kralik (1977), Pollastro (1981), and Diessel (1985), among others. Continued inward alteration and expansion of a biotite stack of crystallites may form a kaolinite vermicule. This is not to imply that all kaolinite vermicules originated as altered biotite stacks; some vermicules in tonsteins seem to have developed instead by crystallization from a gel, possibly growing larger by Ostwald ripening (Eberl et al., 1990). Kaolinitic vermicules of gel origin commonly display central holes (Fig. 12); Hoehne (1964) also reported large vermicules with central holes or "tubes." Many large vermicular stacks seem to be composed of six smaller columns of stacked crystallites arranged around central axis (Plate ID).

Examples of biotite stacks with swollen edges altering to kaolinite coexisting in the same thin section with long,

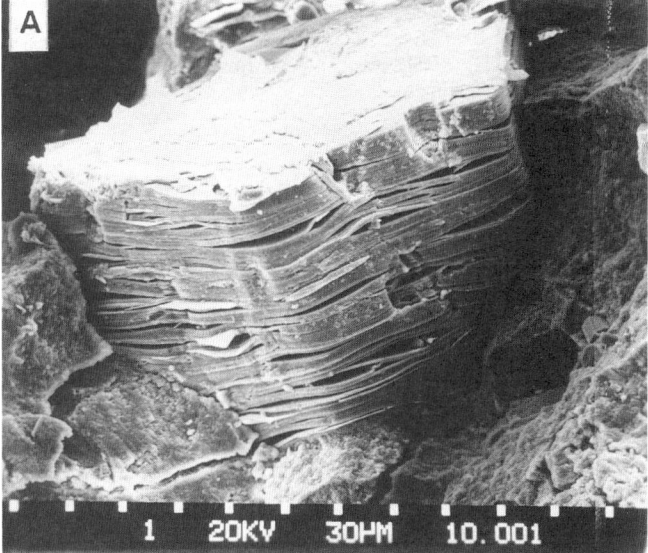
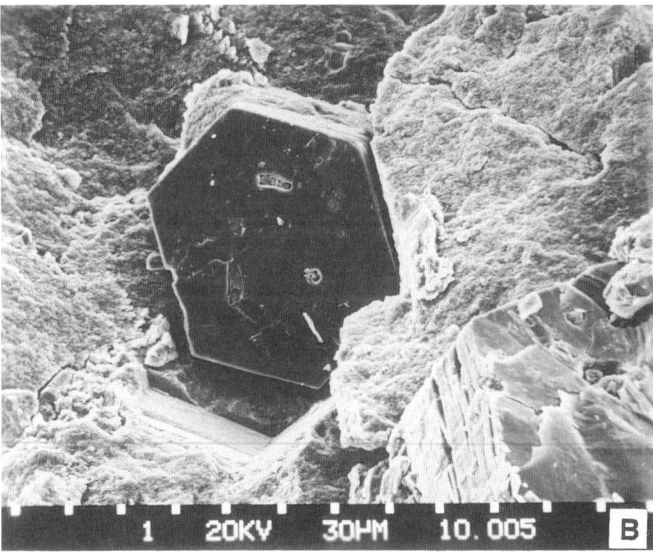
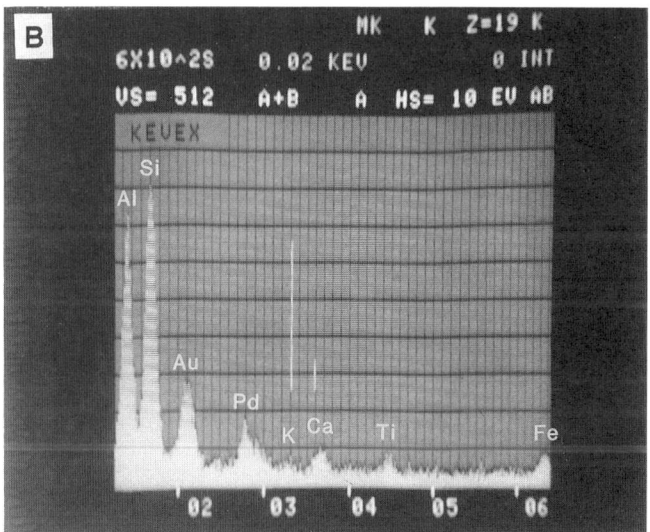

Figure 9. SEM photos of fresh, unaltered euhedral biotite platelets and stacks in tonsteins. A: Platelet with radiation damage hole on front edge, Ferron C coal bed (Upper Cretaceous), central Utah. B: Fresh biotite stack in kaolinite matrix, Denver Formation (Paleocene), Kiowa County, Colorado.

Figure 10. Alteration of biotite to kaolinite in tonsteins. A: SEM photo of biotite stack altering to kaolinite; note disruption of platelets on edge due to alteration. B: Energy-dispersive X-ray spectrum of edge of platelet in A, leached of Fe, Mg, and K (marked by white cursor line). First two peaks on left are Al and Si in ~1:1 ratio, indicative of a kaolinitic composition; third and fourth peaks from left are from from Au/Pd alloy used for coating.

straight-edged, kaolinite vermicules are shown in Plate IE. This occurrence seems to indicate that both modes of genesis are valid; i.e., vermicules can form from the alteration of biotite stacks as well as by direct precipitation. Note that the elongation of the kaolinitic vermicules is normal to the orientation (by sedimentation) of the partially altered biotite stacks in this thin section, cut perpendicular to the bedding plane of the tonstein.

Apatite. Apatite is found in most fresh volcanic ashes of silicic composition. When found in tonsteins, it occurs as euhedral, six-sided, broken prisms due to its strong basal cleavage (Fig. 13). However, apatite is usually not present in most tonsteins because of its susceptibility to acid dissolution. Triplehorn and Bohor (1983) noted the absence of apatite in kaolinite layers (altered volcanic ash) in the upper part of the Dakota Group in Colorado, which was deposited in a nonmarine environment. In contrast, apatite was relatively common in laterally correlative smectitic layers of altered volcanic ash (bentonites) deposited in a marine shale. Triplehorn et al.

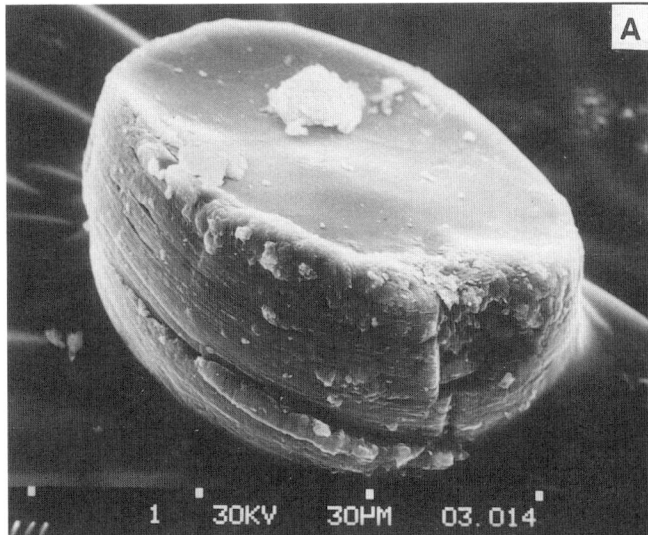

magnetite optically, and hence are grouped together here. Euhedral, six-sided prisms of ilmenite with flat basal terminations are very common in the heavy-mineral fraction of tonsteins (Fig. 14A). Such crystals often have sharp edges, flat crystal faces with bright metallic luster, and show no evidence of weathering or other alteration. Some grains, although bright and shiny, occur as discoids or button shapes without crystal faces (Fig. 14B). This morphology is similar to the rounded droplet-like shapes of resorbed beta-quartz grains noted earlier,

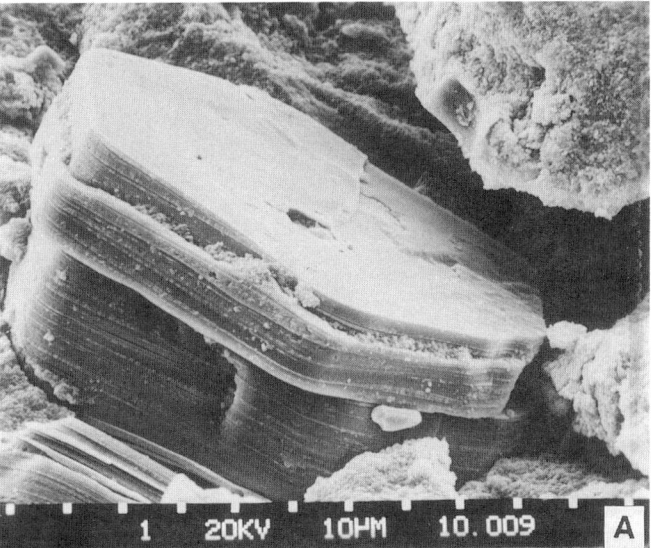

Figure 11. SEM photos of doughnut-shaped altered biotite. A: Biotite stack with swollen outer edges leached of K, Mg, Fe, and Si; now kaolinitic. Core area is still biotite in composition. From middle of thick tonstein, Ferron C coal bed (Upper Cretaceous), central Utah. B: Highly altered and swollen edges of biotite flake; edges are completely kaolinized, while flat central area is still biotitic. From Dakota Formation (Upper Cretaceous), Henrieville, Utah. Compare with Plate IC.

(1977) reported apatite from heavy-mineral separates of Tertiary smectitic tonsteins from Alaska, but these tonsteins were only partially altered and contained fresh glass as well, indicating a low degree of weathering (leaching) by solutions of almost neutral pH. Apatite and other phosphates may also occur as secondary minerals.

Ilmenite and magnetite. These minerals are commonly abundant in tonsteins, occurring as black, opaque, metallic-looking crystals. Ilmenite is not readily distinguishable from

Figure 12. SEM photos showing kaolinite vermicules with hexagonal central holes. A: Vermicular stack of kaolinite plates with central void; note undisturbed layering and lack of swelling of edges. Thick tonstein beyond limits of C coal bed, Ferron Sandstone Member, Mowry Shale (Upper Cretaceous), central Utah. B: Thick plate of kaolinite from vermicule showing hexagonal central hole (in box); Skull Creek Shale (Lower Cretaceous), Colorado.

and may also be the result of magmatic resorption. Anatase and rutile grains displaying an open meshwork texture and pseudomorphing ilmenite crystal shapes are common alteration products of ilmenite (Fig. 15). The meshwork represents exolved lamellae of TiO_2 remaining after solution has removed the interstitial iron-oxide phase (Dimanche and Bartholome, 1976).

Magnetite spherules of several origins have been described by Puffer et al. (1980). They suggest that those of volcanic origin tend to be relatively rich in Ti, whereas extraterrestrial spherules have high Ni contents; manmade (flyash)

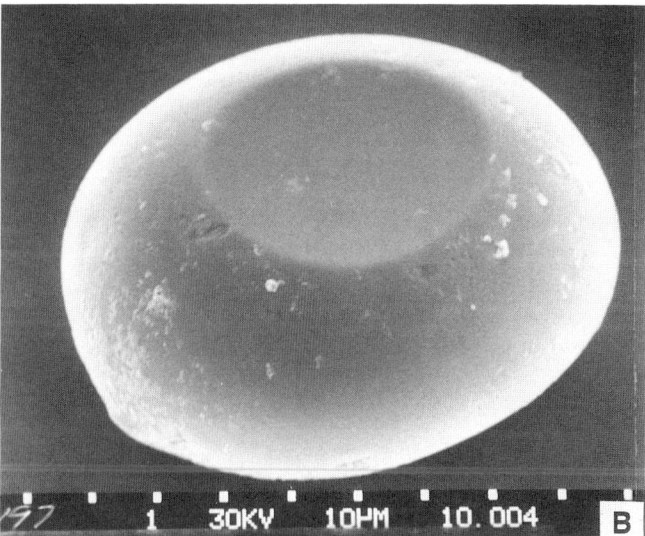

Figure 14. SEM photos of ilmenite crystals. A: Euhedral form with negative crystals in surface, Fire Clay parting, Hazard #4 coal bed (Middle Pennsylvanian), Breathitt Formation, eastern Kentucky. B: Rounded "lozenge" of ilmenite, Ferron C coal bed (Upper Cretaceous), central Utah.

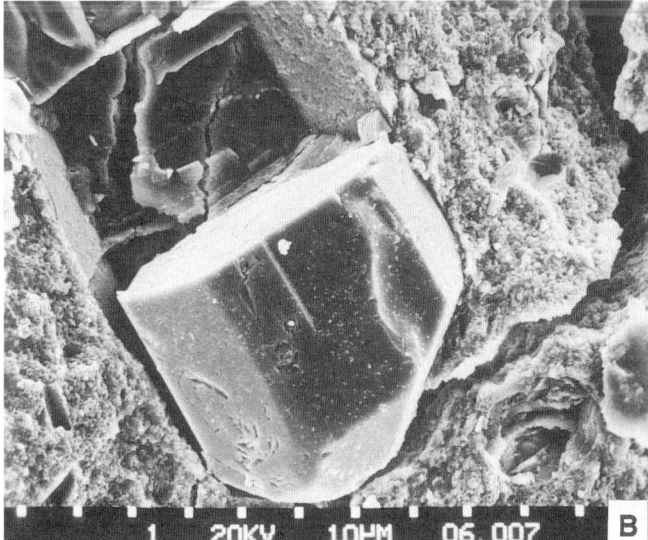

Figure 13. SEM photos of euhedral apatite crystals. A: Crystal with pyramidal termination from center of thick tonstein in C coal bed, Ferron Sandstone Member, Mancos Shale (Upper Cretaceous), central Utah. B: Cleaved prismatic crystal of apatite from thin unnamed coal, upper part of Raton Formation (Paleocene), New Mexico.

spherules lack both Ti and Ni. Volcanic magnetite crystals usually occur as euhedral octahedra.

Zircon. Crystals of zircon are extremely resistant to alteration and are almost always present as euhedra in tonsteins (Fig. 16), even those that have undergone extreme diagenetic modification. Dopita and Kralik (1977) found that a single tonstein is usually characterized by a homogeneous population of euhedral zircons having a limited range of color and shape. However, they found that zircon populations from adjacent detrital sediments tend to be decidedly more variable in color and shape, and are almost invariably rounded to some degree.

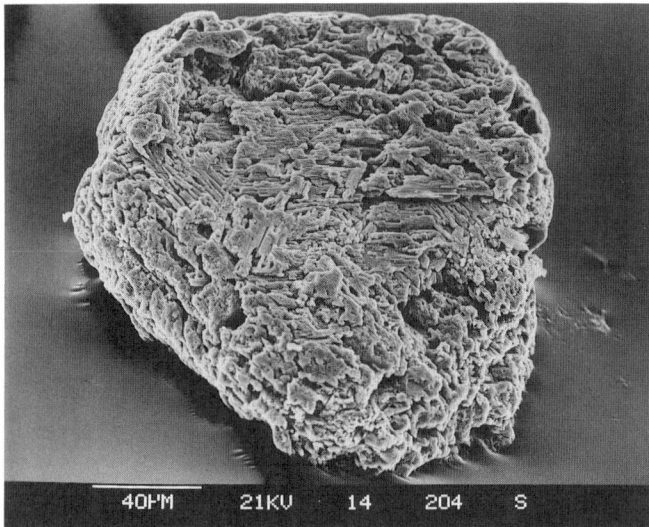

Figure 15. SEM photo of altered ilmenite crystal. Fe has been leached out, resulting in a meshwork of TiO_2 crystals (as anatase) pseudomorphing the original ilmenite crystal. From the Fire Clay parting, Hazard #4 coal bed, (Middle Pennsylvanian), Breathitt Formation, Kentucky.

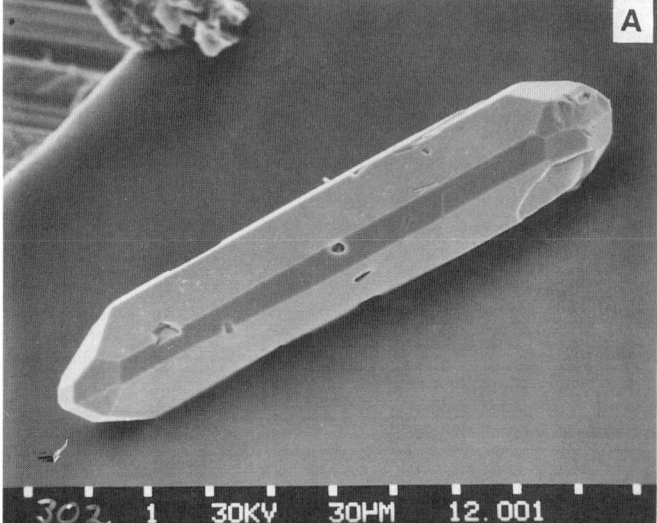

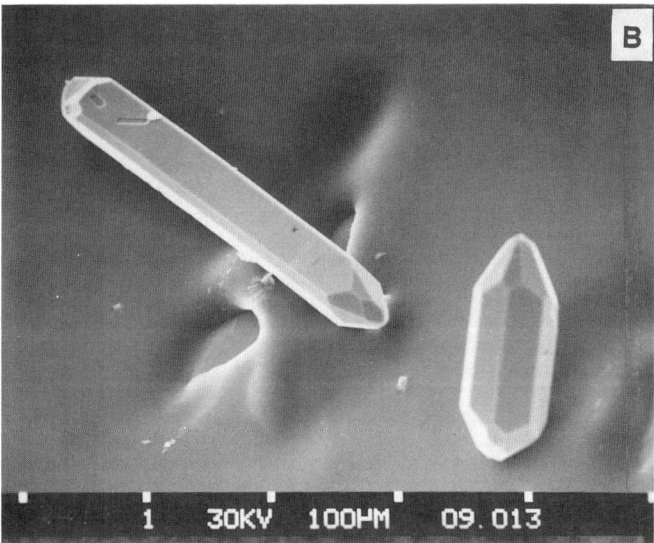

Kowallis and Christiansen (1989) suggested that zircon morphology may be useful for correlation of pyroclastic rocks and applied the classification system of Pupin (1980) to assess the composition and temperature of the parent magma.

Fission-track dating (Triplehorn et al., 1977) and U-Pb isotopic dating (Samson et al., 1989) of zircons from tonsteins (and bentonites) are useful techniques, but caution is necessary because of zoning. Zoned zircon grains may contain cores of older material of different composition that survived one or more episodes of metamorphism and sedimentary recycling.

Rutile and sphene (titanite). These titaniferous minerals can occur in tonsteins, but are rarely abundant; they are also commonly trace components in plutonic igneous rocks. Rutile may alter to leucoxene, which is usually destroyed during the disaggregation and separation processes. TiO_2 may also occur in tonsteins as the minerals brookite or anatase, which are polymorphs of rutile.

Figure 16. SEM photos of euhedral zircon morphologies; note lack of rounding, which is typical of these primary volcanic zircon crystals in tonsteins. A: Elongate zircon crystal from tonstein in I coal bed, Ferron Sandstone Member, Mancos Shale (Upper Cretaceous), central Utah. B: Elongate and stubby euhedral crystals from thick tonstein in C coal bed, also in Ferron Sandstone Member. C: Equant zircon from tonstein in Gallup Sandstone (Upper Cretaceous), Ramah reservation, Zuni, New Mexico.

Other primary components. Glass, allanite, pyroxenes, and amphiboles may survive alteration in younger tonsteins or those that have undergone less diagenesis than older, clay-rich tonsteins. Strongly resistate minerals, such as monazite, topaz, and even garnet and tourmaline, may occur in some tonsteins in trace amounts.

Glass. This is probably the most abundant original component of most ash falls that formed tonsteins. Usually it is totally altered to clay minerals in pre-Tertiary strata, but early cementation by carbonate or silica may preserve unaltered glass in tonsteins (Francis et al., 1968; Jeans et al., 1977). Bouroz et al. (1983) discussed remanent vitroclastic textures, and Burger (1985b, Plate 5) illustrated "volcanogenic glass splinters" in tonsteins. Examples of glass textures and relict shards were shown by Burger (1990) and Burger and Wolf (1987). Although glass shards have frequently been described in tonsteins, it is not always clear whether the author actually detected true glass or only shard-like clay mineral pseudomorphs. We have observed glass bubble junctions and shards replaced by both kaolinite and alumino-phosphate (Fig. 17, A and B). Glass morphology and composition in volcanic ashes have been well documented by Fisher and Schmincke (1984) and Heiken and Wohletz (1985).

Allanite, pyroxenes, and amphiboles. Allanite, a member of the epidote group rich in rare-earth elements (REE), has been reported in tonsteins from Eocene coal beds in Wyoming (Bohor et al., 1979). Allanite is easily altered; it is only rarely found in tonsteins in trace amounts as corroded crystals. Hornblende, however, is sometimes relatively abundant in partially altered tonsteins and can be used for radiometric dating (Triplehorn et al., 1977). Most pyroxenes and amphiboles are present only in trace amounts in silicic volcanic ashes, but are common components in intermediate to basic ashes (Sanchez et al., 1987); usually these minerals are readily altered in the coal-forming environment.

Trace heavy minerals. Weaver (1963) emphasized the interpretive value of heavy minerals from bentonites for determining whether they were primary volcanic, secondary (redeposited) volcanic, or nonvolcanic. He stated that primary volcanic bentonites should have a heavy mineral suite restricted to biotite and idiomorphic zircon, apatite, and titanite, although younger, less-altered bentonites also can contain hornblende, augite, and hypersthene. Weaver designated bentonites containing toumaline, garnet, and muscovite (and apparently rutile) as secondary (reworked and contaminated); however, some of his own data seem to contradict these conclusions. Literature reports of rhyolitic lavas and pyroclastics containing tourmaline, pink garnet, and topaz suggest that these minerals can indicate a primary volcanic origin when found in tonsteins or bentonites, especially if they are idiomorphic and occur in only trace amounts. However, muscovite and the diagnostic metamorphic minerals kyanite, sillimanite, staurolite, glaucophane, and chloritoid are sure indicators of redeposition and contamination. It seem clear that a judgement of origin for

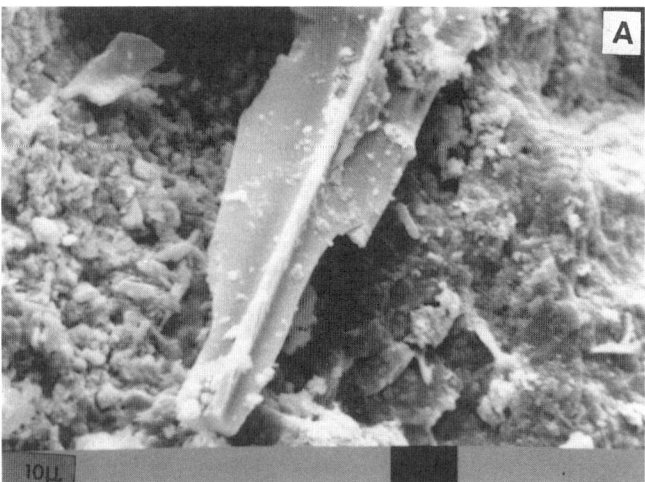

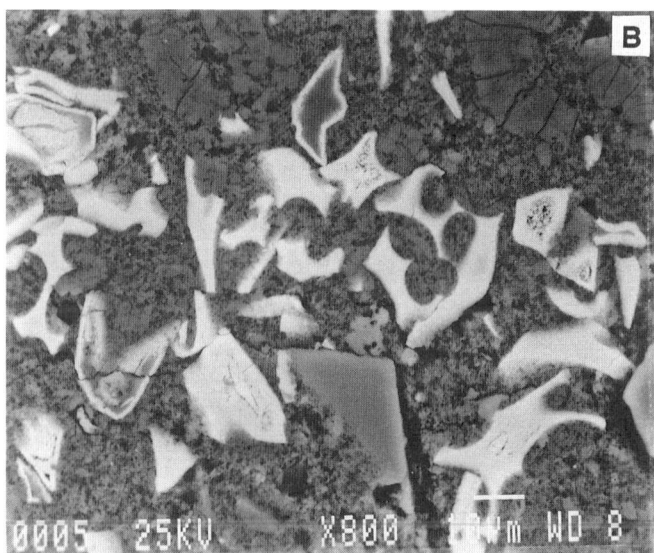

Figure 17. Pseudomorphed glass in tonsteins. A: Partial bubble junction replaced by kaolinite; from Felix coal bed, Wasatch Formation (Eocene), Powder River basin, Wyoming (width of lower tip of shard is 4 µm; SEM photo. B: Glass shards and bubble junctions (white) replaced by alumino-phosphate (crandallite); from thin tonstein in the Cannal City coal bed (Middle Pennsylvanian), Breathitt Formation, Kentucky. SEM photo (back-scatter mode); scale bar = 10 µm.

clays such as tonsteins needs to be based on more than just the heavy mineral suite. Spears (1982) also expressed doubts about the absolute value of heavy minerals as genetic indicators in clays, and stated that heavy minerals may or may not be diagnostic of volcanic origin, depending on the presence or absence of normal clastic minerals.

Secondary Minerals

Kaolinite. Just as smectite is the characteristic clay mineral of most bentonites, kaolinite is the most common and abundant clay mineral in tonsteins; however, exceptions occur (Triplehorn et al., 1977). Kaolinite forms mainly from the alteration of glass, but feldspars, amphiboles, pyroxenes, and other minerals also alter to kaolinite. Kaolinite may be pseudomorphic after feldspar, glass, biotite (as noted previously), or ferromagnesian minerals. Cryptocrystalline kaolinite may also precipitate within plant cells or root traces.

X.ray diffraction analyses of tonsteins typically show well-crystallized kaolinite as the dominant or sole component of the clay fraction (Fig. 18). The purity and high degree of crystallinity of the kaolinite shown in these analyses are evidence for its authigenic origin.

Kaolinite commonly occurs in tonsteins as large platey crystals or in elongate, vermicular stacks of crystals with a well-developed hexagonal outline (Fig. 19, A and B). Crystals of kaolinite only a few microns in size may occur as pore-filling clay or dispersed in a microcrystalline matrix. The largest vermicules may attain lengths of several millimeters; these are often visible to the naked eye. Vermicular kaolinite is not confined to tonsteins, but some of the largest and best-formed examples may be found in these beds. The presence of visible kaolinite vermicules often aids in the field identification of tonsteins.

Smectite. This mineral commonly occurs in tonsteins whose parent ash was intermediate in composition. For example, smectite is common in tonsteins of the Skookumchuck Formation of western Washington (Turner et al., 1983;

Figure 19. SEM photos of coarse crystalline and vermicular kaolinite. A: Large crystals and platelets of kaolinite; Upper Permian coal, Yunnan Province, China. B: Large curved vermicules, Felix coal bed, Wasatch Formation (Eocene), Powder River basin, Wyoming.

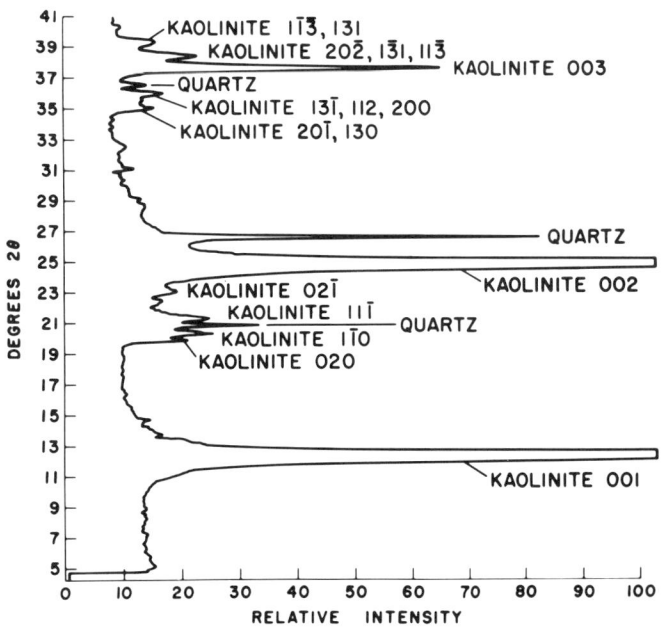

Figure 18. X-ray diffraction pattern of typical tonstein showing sharp basal peaks and prism reflections of kaolinite, indicative of authigenic origin. Dawson Formation (Paleocene), Denver basin, Colorado.

Reinink-Smith, 1982). Smectite is also commonly reported as occurring within the central portion of thick tonsteins (Triplehorn and Bohor, 1981; Bouroz et al., 1983; Zaritsky, 1985). The presence of smectite here reflects incomplete alteration due to a lesser degree of flushing (leaching) of the interior of the bed, in contrast to the kaolinitic marginal portions of these tonsteins; thin tonstein layers in the same coal-bearing intervals are usually composed completely of kaolinite. Smectite-bearing tonsteins are also characterized by lesser alteration of labile phenocrysts of biotite, ferromagnesian minerals, and feldspar.

Illite. This mineral has rarely been reported in tonsteins of Cretaceous and younger ages in the United States. However, both discrete illite and illite/smectite (I/S) have been reported from some Carboniferous tonsteins of Europe (Stöffler, 1963; Spears, 1971; Dopita and Kralik, 1977; Burger and Stadler, 1984), and from Permian tonsteins in Australia (Kisch, 1966) and China (Burger et al., 1990). Bouroz et al. (1983) related its presence to original ash composition and depth of water in the depositional environment. However, this model is not universally applicable because no illite has been reported in tonsteins from similar depositional settings in western North America. Zaritsky (1985) found illite occurring with smectite in tonsteins from the Soviet Union and attributed this association to incomplete leaching. Discrete illite is often found interlaminated with kaolinite in vermicules, indicating either that the vermicules represent altered biotite stacks or that these particular layers represent illitization of kaolinite. Bouroz et al. (1983) showed that illite can pseudomorph glass shards, and suggested that it may also replace quartz. Illite and I/S in older tonsteins probably is formed from smectite by burial diagenesis (Pollastro, 1983; Burger et al., 1990).

Phosphates. Phosphate minerals have practical and geologic significance for both coal and tonsteins. High phosphorous content may reduce the economic value of coal for coking. Out of a suite of 31 European tonsteins, the mean P_2O_5 content in 10 was 4.36% (Spears and Kanaris-Sotiriou, 1979, Table 2), and Reinink-Smith (1990a) reported >28% P_2O_5 in an Alaskan tonstein. These phosphate contents were ascribed to alumino-phosphate minerals.

As noted earlier, phenocrysts of apatite occur in trace amounts only rarely in tonsteins, although it is a common phenocrystic component in bentonites; this mineral is not stable in an acidic environment. Minor amounts of secondary apatite also occur in tonsteins as microcrystalline cement or replacing primary minerals (Addison et al., 1983; Strauss, 1971; Hoehne, 1959; Dopita and Kralik, 1977). Secondary apatite often occurs along with unaltered primary apatite crystals, indicating that the secondary apatite is probably not derived from solution of primary crystals.

Alumino-phosphate minerals of the goyazite-crandallite group (Fig. 20) have long been recognized as trace components of tonsteins, but their distribution and exact compositions were obscure in earlier studies (e.g., Stadler and Werner, 1962; Burger, 1964; Price and Duff, 1969). The major cations determine the end-member species within a solid-solution series: Ca (crandallite), Sr (goyazite), Ba (gorceixite), and Ce and other rare earths (florencite). Ca- and Sr-rich types are common, but most occurrences are mixtures of some or all of the major species noted above. Identification is usually made initially by X-ray diffraction (Fig. 21). Our North American studies (Triplehorn and Bohor, 1983) suggest that such minerals may be more widespread in tonsteins than has generally been recognized, and sometimes may even be the dominant component (Reinink-Smith, 1990a). Triplehorn and Finkelman (1989) found crandallite pseudomorphic after glass shards and bubble junctions (Fig. 17B). These occurrences suggest that phosphate replacement occurred soon after burial, and before the glass could be altered to clay minerals by diagenesis.

Figure 20. Goyazite (Sr-rich member of crandellite series of alumino-phosphates). A: Concentrate (HF-insoluble residue) of subhedral to anhedral grains. B: Euhedral crystals precipitated as cavity lining, showing rhombohedral (pseudocubic) habit. SEM photos; numerical scales at bottom edges indicate micrometer distances between tick marks. Kaolinitic bentonite in Mowry Shale (Upper Cretaceous), Morrison, Colorado.

Alumino-phosphate minerals are extremely insoluble and high in density, allowing them to be concentrated by removal of silicates with HF (Norrish, 1968), followed by panning or heavy-liquid separation. If the microcrystalline aggregates are dispersed, they can easily be lost during decantation of the fine fraction. The lack of relatively large single crystals or pure concentrates of fine aggregates of these minerals makes it difficult

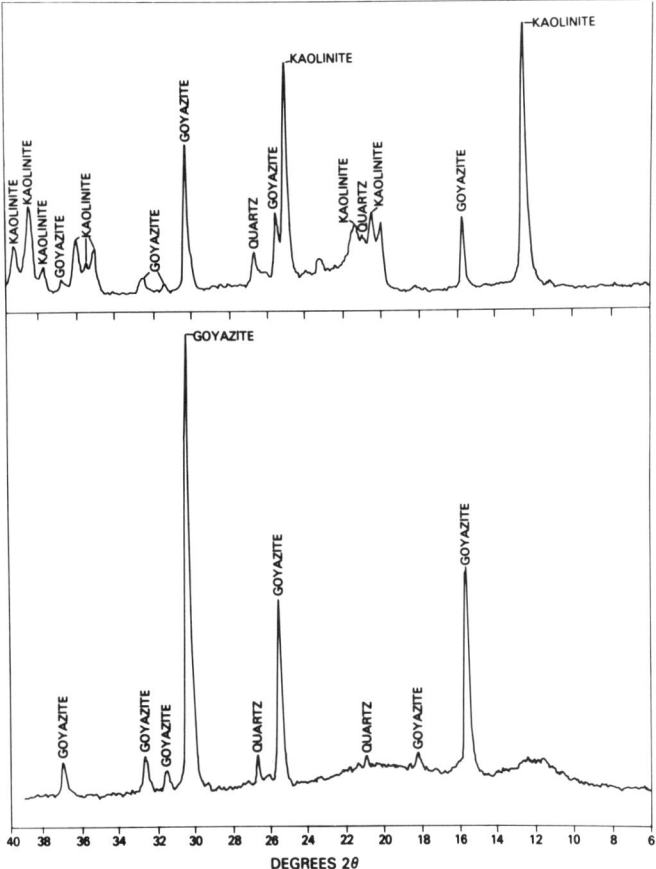

Figure 21. Diagram showing goyazite X-ray diffraction peaks in a typical tonstein powder pattern (upper), and as a purified goyazite separate (lower). Kaolinitic bentonite in Mowry Shale (Upper Cretaceous), Morrison, Colorado.

to determine their exact chemical or mineralogical composition.

The origin of secondary phosphate minerals has received little attention. The source of the phosphorous and the time and place of crystallization are not well known. As noted previously, the original volcanic ash is not an important source of phosphorous, which lead Spears et al. (1988) to suggest an alternate source from organic material. Wilson et al. (1966) made a similar suggestion and pointed out that up to 7.4% P_2O_5 had been reported in the ash of a modern lycopod. Although details are not entirely clear, studies on phosphorous behavior in soils developed on pyroclastics (andosols) may provide a good model for the origin of secondary phosphate minerals in tonsteins (see Tan, 1984). Under acid conditions, organic phosphorous in soil water reacts rapidly with weathered glass (allophane and alumina) to form insoluble aluminum phosphates; these then gradually change to a stable crystalline phase. The accumulation of phosphorous is so rapid and so linear with time that it has been suggested that phosphorous content might be used to estimate the age of soils up to about 8,000 years old (see Tan, 1984). Thus, secondary phosphate mineralization may also provide insights on early diagenesis in tonsteins.

U and Th in phosphates may be the source of the radioactivity of tonsteins noted on gamma-ray logs by several workers (e.g., Williamson, 1970a, 1970b; Zaritsky, 1971; Dopita and Kralik, 1977). The radioactive response of some tonsteins on gamma-ray logs enhances their utility as isochronous marker beds for correlation. It should be noted, however, that some tonsteins with low P_2O_5 contents also appear to be radioactive (Spears, 1966, 1971; Spears and Rice, 1973; Spears and Kanaris-Sotiriou, 1979). In such cases, these authors attributed the gamma response to U in zircon and to Th in kaolinite; other primary volcanic minerals could also contribute to the radioactivity of tonsteins.

Pyrite. Authigenic pyrite is common and locally abundant in tonsteins. It occurs in several forms normally found in coal: euhedral cubic and octahedral crystals, framboids, fillings of cell lumens and other cavities, and as replacements of organic materials.

Diatoms. Siliceous diatom tests are common components of some Tertiary coal beds and associated tonsteins (Triplehorn, 1976a; Sanchez et al., 1987). Because opaline silica is unstable, diatoms commonly dissolve during the diagenesis of tonsteins. However, diatoms are relatively more stable than volcanic glass, because they coexist with completely altered glass in some tonsteins. Volcanic ash can provide the silica for diatom growth, and diatom "blooms" are often associated with ash deposits. Diatoms are restricted to rocks of Jurassic and younger age.

Weathering products. Minerals such as pyrite and other sulfides, ferrous compounds, and carbonates may produce a large variety and abundance of secondary minerals upon weathering. Most common are jarosite, gypsum, barite, and hydrated iron oxides; calcite and melanterite are less common. Generally, any of the weathering products associated with coal can also occur in tonsteins.

GROSS TEXTURE OF TONSTEINS

The texture of tonsteins, as seen in thin section, has been the primary basis for their classification. Bouroz et al. (1983) and Skocek (1973) provided a history and discussed these schemes; most are based on the texture of neoformed kaolinite minerals and replaced pyroclastic components.

Coarse-grained tonsteins have often been mistaken for sandstones during field examination. Large plates and vermicules of kaolinite stained with brownish organic coatings lend the appearance of a coarse-grained sandstone to these layers. Close examination with a hand lens should reveal the typical hexagonal shape of the kaolinite forming these grains, reducing the need for laboratory analysis for their identification. Specific textures and bedding relations also may be helpful in identifying the origin and diagenesis of tonsteins.

Plate I (on following three pages). A: Crystal tonstein with extremely high ratio of primary volcanic crystals to clay matrix; crossed polarizers with gypsum plate. "Big Dirty" coal bed, Fort Union Formation (Paleocene), near Roundup, Montana (see back cover photo). B: Beta-form quartz paramorph crystals with glass inclusions. Cross-polarized light and oblique illumination in immersion oil; average crystal diameter = 190 µm. Flint clay marker (noncoal tonstein) from underclay at base of the Princess #6 coal zone, Breathitt Formation (Middle Pennsylvanian), near Louisa, Kentucky (see Outerbridge et al., 1990). C: Layer of biotite packets (crossed polarizers with gypsum plate) aligned normal to bedding with swollen, fan-shaped edges altering to kaolinite. Comanche coal bed, Denver Formation (Paleocene), Elbert County, Colorado. D: Long, straight-sided kaolinite vermicule (crossed polarizers) with growth axis parallel to bedding, probably formed by crystallization from solution or gel. Unknown Paleocene coal bed (from core), Watkins, Colorado, Denver basin. E: Thin-section photo (crossed polarizers with gypsum plate) of tonstein cut perpendicular to bedding, showing edge-expanded biotite packets normal to bedding (top half) and kaolinite vermicules with long axes parallel to bedding (lower half). Thus, both types of vermicular development can occur within the same thin section. Same tonstein and location as D. F: Graupen (red) containing crystals of quartz and feldspar (white, yellow, and blue). Thin-section photo (crossed polarizers with gypsum plate) of tonstein in Denver Formation (Paleocene) coal bed, Elbert County, Colorado.

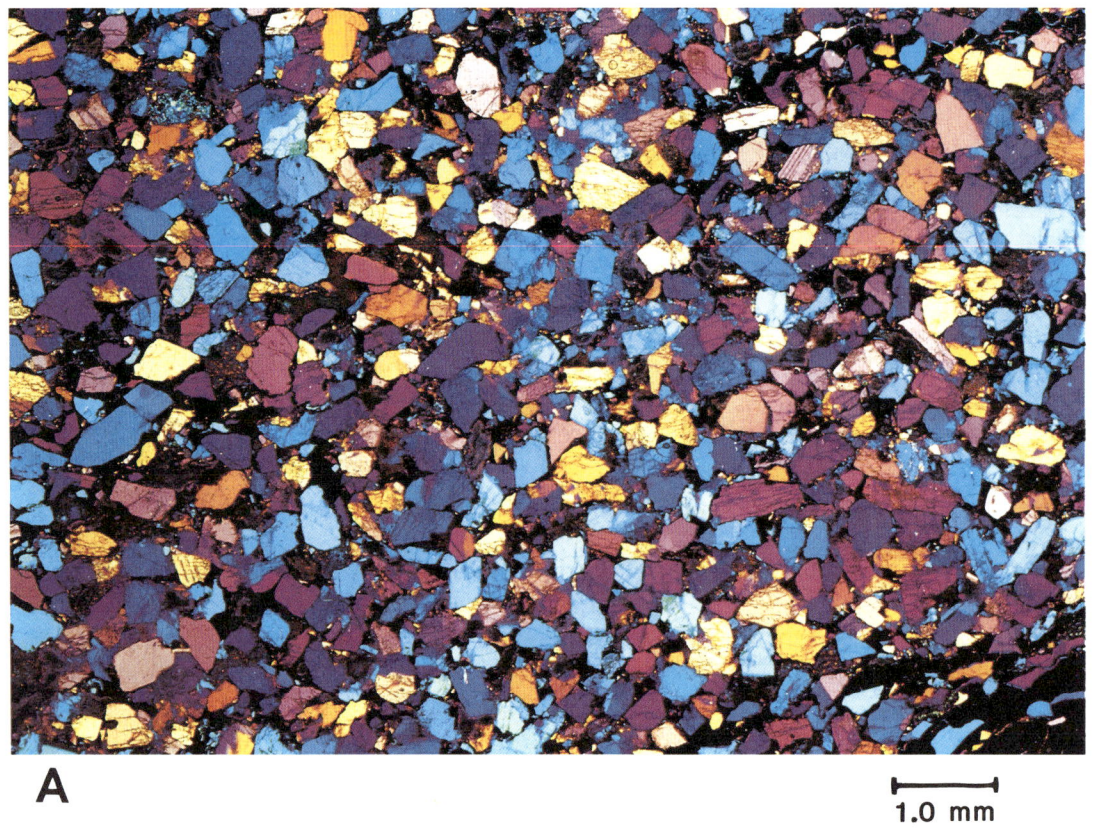

A

B

C

|—————|
0.1 mm

D

|————————————|
1.0 mm

E

|—————| 1.0 mm

F

|—————| 1.0 mm

Vermicules, plates, and microspherules

Kaolinite commonly occurs in tonsteins as elongate, curved stacks of pseudohexagonal platelets called vermicules (Fig. 22, A and B), that can grow to a relatively large size (1–5 mm). Stacks of large kaolinite platelets can be formed by replacement of biotite stacks, which expand in the alteration process. Some vermicules may form in this way, but others are clearly chemically precipitated, without any obvious biotite precursor. In either case, solution and short-range transport of material is indicated, including removal of much of the silica. The fragile nature of these vermicules indicates an authigenic origin during diagenesis, because they could not have survived much transport.

A relatively rare texture in kaolinitic tonsteins is best described as microspherulitic. Examples of this texture were shown by Dopita and Kralik (1977, p. 205, Table XXXV). We have also observed this texture occasionally, specifically in tonsteins from the Denver basin, Colorado. The morphology of these tiny microspheres closely resembles spherical halloysite, probably developed from allophane, that has altered to kaolinite. A similar texture of microspherical kaolinite "honeycomb" has also been observed in the basal kaolinitic layer of altered impact ejecta at Cretaceous-Tertiary (K/T) boundary sites in the Western Interior (Bohor, 1983; Pollastro et al., 1983). This texture in K/T ejecta is illustrated by Pollastro and Pillmore (1987). Both the volcanic-ash precursor of the tonsteins and the impact glass of the K/T boundary basal impact ejecta probably were siliceous, and both air-fall deposits altered under similar conditions in coal-accumulating environments (Pollastro and Bohor, 1992).

Accretionary lapilli

These features, well known from recent volcanic tephra deposits, were first described in European cinerites by Bouroz et al. (1983) and in North American tonsteins by Bohor and Triplehorn (1984). Accretionary lapilli are spheroidal (usually oblate spheroids; Fig. 23A) bodies (<1 mm to ~8 mm in maximum diameter in tonsteins) and resemble oolites because of their concentrically layered outer rims, but lack any strong mineralogical contrast between their cores and the matrix of the enclosing tonstein (Fig. 23B). Accretionary lapilli and their origin in proximal pyroclastic deposits are well described in the literature (e.g., Moore and Peck, 1962; Bateson, 1965; Schumacher, 1988). In tonsteins, accretionary lapilli indicate a pyroclastic origin; they formed in relatively dense distal ash clouds where moisture was abundant, perhaps forming during thunderstorms (Bohor and Triplehorn, 1984).

In flint clays and other intensely leached, high-alumina clays, the occurrence of accretionary lapilli is still controversial. Such clays often develop pisolitic structures that resemble accretionary lapilli. Bohor and Triplehorn (1984) discussed criteria for distinguishing between these pisolites and accretionary lapilli, but a clear distinction is not always possible.

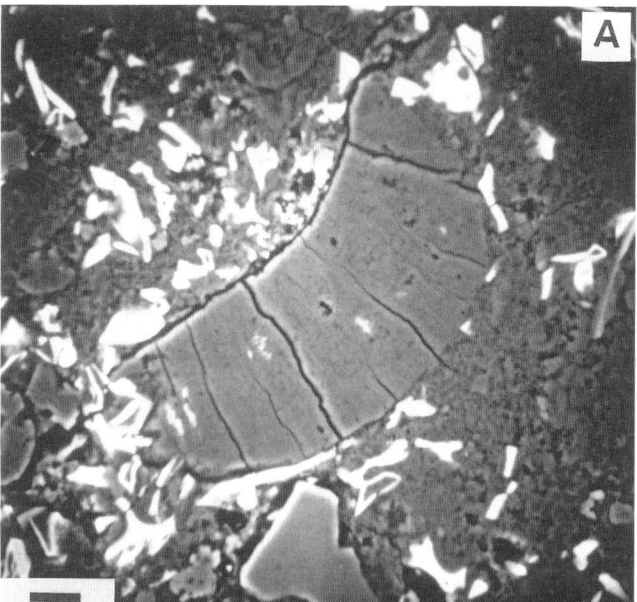

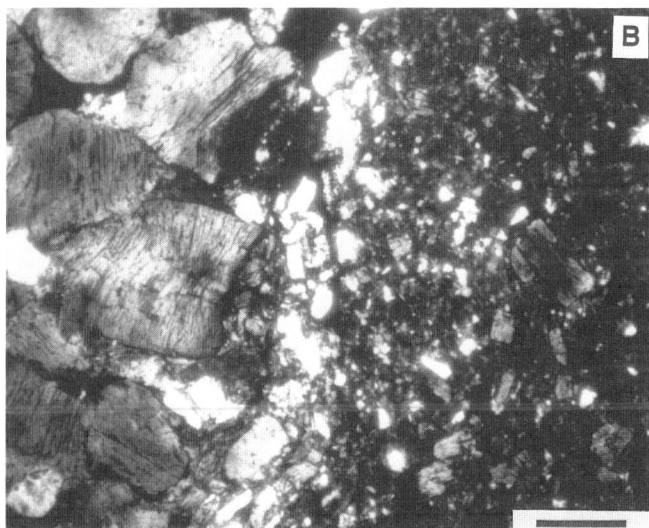

Figure 22. Kaolinite vermicules in matrix. A: Curved vermicule of kaolinite (light gray) surounded by glass shards (white) replaced by aluminophosphate (crandallite). Note how growing vermicule has displaced glass shards and accumulated them along its borders. Cannal city coal bed (Middle Pennsylvanian), Breathitt Formation, eastern Kentucky; SEM photo in back-scatter mode; scale bar = 10 µm. B: Thin-section photo of large swollen vermicules (left), white quartz splinters displaced by growing vermicules (middle), and finer kaolinitic dark matrix (right); plane-polarized light, scale bar = 100 µm. Lower part of Denver Formation (Paleocene), Denver basin, Colorado.

Graupen

Ovoidal or ellipsoidal aggregates of microcrystalline kaolinite within a coarser matrix are called graupen (from the

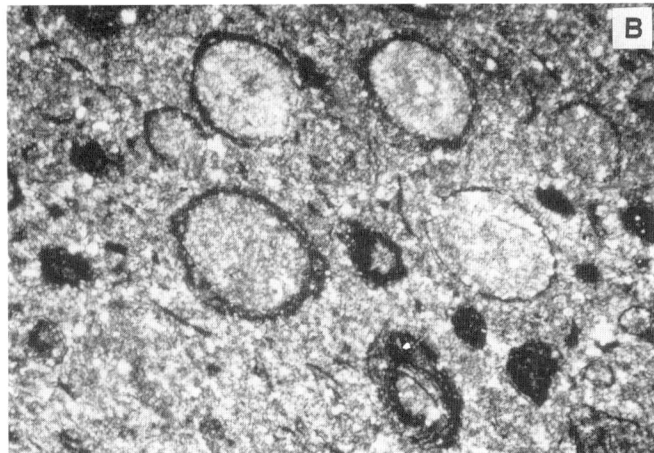

Figure 23. Accretionary lapilli in tonsteins. A: Spherical white accretionary lapilli weathering out of altered silicified volcanic ash; Frontier Formation (Upper Cretaceous), Gros Ventre River, Wyoming; photo of weathered natural surface (plane-polarized light); white spherules range between 3–5 mm in size. B: Accretionary lappili in thin section; note broken rims and rim fragments (dark) and similarity of composition, grain size, and texture between lapilli cores and matrix (plane-polarized light). Flint clay in Olive Hill Clay Bed (Lower Pennsylvanian), Lee Formation, Clack Mountain locality, Kentucky.

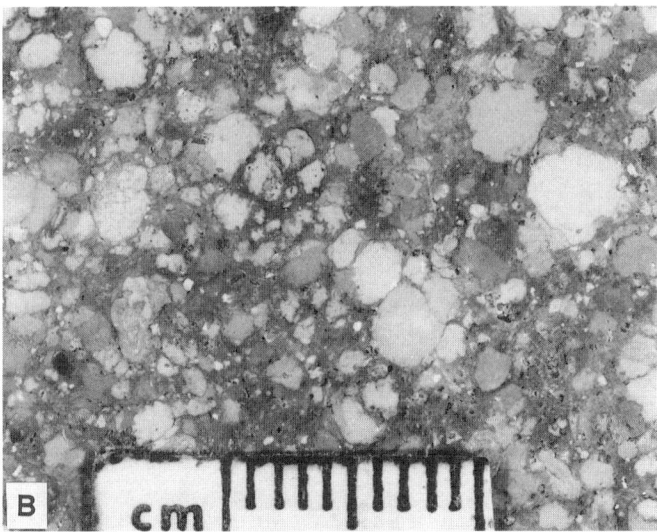

Figure 24. Graupen textures in tonsteins. A: Cross-sectional view of graupen. B: Plan view of graupen. C coal bed, Ferron Sandstone Member, Mancos Shale (Upper Cretaceous), central Utah. Ordinary light photographs of polished blocks; scale divisions = 0.1 cm.

German word for "grapes"). They occur as large as a few millimeters in diameter and are often visible to the unaided eye in polished specimens (Fig. 24, A and B). Graupen abundance ranges from scattered individuals to a major portion of some tonsteins. The abundance and size of graupen commonly change vertically in tonsteins, suggesting a gradation that may have been determined either by the original size and distribution of these clasts in the ash cloud, or by flotation and settling in water. Graupen are often concentrated in zones at the tops and bottoms of tonsteins layers (Dopita and Kralik, 1977).

The origin of graupen is still unknown. A major question is whether they grew as purely authigenic aggregates, or as replacements of some primary textural component, such as pumice; perhaps they have more than one origin. Graupen are typically composed of fine-grained kaolinite, and no primary textures are preserved that would help explain their origin. Smooth, ovoid-shaped graupen composed of pure fine-grained kaolinite may represent infilled bubble shards (Fisher and Schmincke, 1984, p. 102). Most of the typical graupen that we have observed in tonsteins resemble tiny pumice lumps that may have been filled with authigenic kaolinite by a process of concurrent glass solution and clay precipitation. Thin-section photos (Plate IF) reveal primary volcanic mineral grains (mostly quartz) within some of the graupen, reinforcing the pumice theory of origin. Light-colored areas in this photo may represent void-filling precipitated kaolinite, whereas dark-colored areas may indicate glass replacement by kaolinite. Observations that graupen zones are often confined to tops and bottoms of subaqueously deposited tonsteins also suggest a pumice origin, because some pumice clasts sink and others float (Fisher and Schmincke, 1984, p. 117).

Breccia

Tonsteins and other fine-grained kaolins occasionally display brecciated textures. The term "fragmental clay rocks" (FRC) has been applied to these brecciated claystones (Richardson and Francis, 1971). The origin of this brecciated texture is problematic; several processes may be involved. Where rounding, sorting, or other evidence of current transport are present, weathering and erosion can be invoked (Loughnan, 1978). More puzzling are cases that show no such erosional evidence, particularly those involving thin tonsteins within coal beds. Although the mechanism is not entirely clear, brecciated textures can be related to desiccation or shrinkage, with little or no transport of the disrupted fragments (Richardson and Francis, 1971). The texture of these FRCs may be internally generated by postdepositional processes unrelated to normal weathering and erosion, such as those operating in the formation of intraformational breccias and in pedogenesis.

Flint clay

"Flint clay" is used here as a field term for microcrystalline or cryptocrystalline clays that lack lamination, break with conchoidal fracture into angular fragments, and resist slaking in water. Laboratory studies show that they are composed of pure, well-crystallized kaolinite. Some thick flint clays are not tonsteins in the sense of having a direct ash-fall origin, and most tonsteins do not have the texture of flint clay. Thin tonsteins with the characteristics of flint clays can occur in coal beds, however. The Fire Clay parting (tonstein) in the Hazard #4 coal bed of Kentucky has a flint-clay texture (Seiders, 1965; Bohor and Triplehorn, 1981; Triplehorn et al., 1989), and other examples of tonsteins with this texture have been found in the Matanuska Valley of Alaska (Triplehorn, unpublished data). In these cases, primary minerals, textural components, and field relations clearly indicate an air-fall volcanic-ash origin for these thin layers. However, thick flint-clay beds often may not display clear evidence of air-fall origin; in the past, these clays have been assigned origins related to intensive leaching of swamp sediments or adjacent soils. However, portions of the thick Olive Hill Clay Bed, a flint clay in eastern Kentucky, contain evidence of an ash-fall origin (Bohor and Triplehorn, 1984), including euhedral beta-quartz crystals (Fig. 25, A and B). This mineralogical evidence contravenes the the-

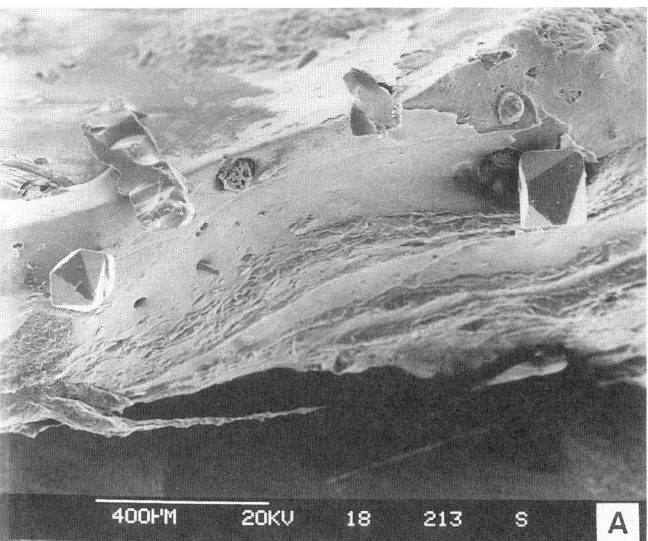

Figure 25. Phenocrysts of quartz etched out of tonstein surface with HF. SEM photos. A: Two beta-form quartz paramorph crystals and an irregular quartz splinter (upper left) etched into relief from matrix of fine-grained kaolinite. B: Closeup of beta-form quartz crystal etched free of matrix. Flint clay from Olive Hill Clay Bed (Lower Pennsylvanian), Lee Formation, Kentucky.

ory that this flint clay was produced entirely by the weathering (alteration) of ordinary detrital sediments in a coal-forming environment (Patterson and Hosterman, 1958). Other flint clays may also show evidence of a possible volcanic heritage if subjected to a detailed mineralogical analysis of their nonclay mineral components.

Clay-free partings

Many thick coal beds do not contain visible tonstein partings, even at stratigraphic positions where such partings would normally be expected to occur (Francis, 1985). Discrete, clay-free partings, consisting of individual grains (usually no more than a single monolayer) of quartz, feldspar, and other primary volcanic resistate minerals scattered on a bedding surface, have only recently been identified in both thick and thin coal beds. Some glass must have originally accompanied these volcanic resistate mineral grains, because it is impossible for a parent magma to have consisted solely of phenocrysts. However, there is no evidence of original glass or labile minerals present in the form of clay-mineral replacements or pseudomorphs in these partings. We have observed these clay-free volcanic partings in coal beds from Montana, the Powder River basin of Wyoming, the Denver basin of Colorado, and the Cook Inlet and Nenana basins of Alaska. Only the Powder River basin occurrence in the Wyodak-Anderson coal bed has been documented (Triplehorn et al., 1991).

Clay-free volcanic partings represent a special case where the glass phase is dissolved and removed from the system, rather than being altered to, or precipitated as, clay minerals. Because these volcanic ash partings do not alter to claystones, they cannot be designated as tonsteins. The term "cinerite" was proposed (Bouroz, 1962) for all ancient pyroclastic layers regardless of their degree of alteration or present mineral composition; therefore, these clay-free partings can only be ultimately classified as cinerites.

The absence of clay minerals due to the solution of glassy and other labile phases implies a substantial loss of silica and alumina from the system, although alumina has been considered to be relatively immobile during weather (Spears and Kanaris-Sotiriou, 1975). However, even such immobile constituents as titanium often show evidence of short-range transport in tonsteins, being codeposited with silica and kaolinite as fillings in plant cells (Triplehorn et al., 1991).

Clay-free partings are significant because of their potential for correlation, age dating, and environmental interpretations; however, they are easily overlooked by investigators unaware of their possible existence in thick coal beds. The intense leaching and solution transport implied by the absence of clay minerals in these partings suggests a possible relation to the raised-bog model of coal formation (cf. McCabe, 1984). The very thick Wyodak-Anderson coal bed in the Powder River basin lacks visible tonstein layers, whereas clay-rich tonstein partings are present in most thinner coal beds (planar peat model) of this region. A raised-bog model has been proposed for the Wyodak-Anderson coal bed (Warwick and Stanton, 1988). According to this model, a chemical environment is produced in raised (ombrogenous) bogs that is particularly conducive to solution of labile components, rather than their alteration to clay minerals as predicated by alternative models of coal formation (Triplehorn et al., 1991). Probably only in thin ash-fall layers would the glass and labile minerals completely dissolve under ombrogeous conditions.

Glass shards and lapilli

Sometimes original pyroclastic particle morphologies are preserved in tonsteins, most often pseudomorphed by kaolinite or other minerals. Bohor and Pillmore (1976) reported kaolinite pseudomorphs of glass bubble junctions in a tonstein from the Felix coal bed (Eocene) of Wyoming (Fig. 17A). Several instances of glass shards, spherules, or droplets replaced by alumino-phosphate minerals of the crandallite group have been reported in tonsteins from a wide range of ages and localities (Triplehorn et al., 1991; Triplehorn and Finkleman, 1989). Kaolinized lapilli with aerodynamic shapes and delicate tails (Fig. 26, A, B, and C) represent volcanic-glass droplets replaced during diagenesis of the tuff to a tonstein. Bouroz et al. (1983) have also reported illite pseudomorphs after glass shards in tonsteins. These replacements of large glass clasts by clay minerals probably required a greater period of time after deposition than did the solution of the finer glass particles (dust) and the precipitation of kaolinite in the matrix.

Rooting and burrowing

Resumption of plant growth following an ash fall often results in root penetration of the nascent tonstein. Larger roots are usually carbonized and are often visible in outcrop. Less conspicuous are the smaller rootlets; these may be abundant, but are only observable on polished surfaces or in thin section. Some rootlets are preserved as casts of cryptocrystalline kaolinite, implying early decay of the organic material and rapid filling by chemically precipitated kaolinite.

Burrows are rarely identified in tonsteins; organisms active in fresh water are usually responsible for those that are found. However, Triplehorn and Bohor (1981) described burrows of marine organisms in tonsteins from the Cretaceous Ferron Sandstone Member of the Mancos Shale in Utah. In this occurrence, the burrowing took place only after exhumation of the partially lithified tonstein by marine erosion. Sand similar to that forming the overlying marine sandstone fills vertical burrows in the underlying tonstein. This evidence shows that the tuff precursor of the tonstein began to consolidate and lithify soon after deposition, resisting erosion by the marine incursion and yet remaining soft enough to be penetrated from above by burrowing marine organisms.

Stumps, twigs, and other plant materials

Preservation of large plants in growth position is uncommon in tonsteins, even though deposition of air-fall material ideally should preserve most details of the accumulation surface. Perhaps this is a matter of preferential preservation, in that abundant standing vegetation may prevent the formation of sharply defined, thin ash layers.

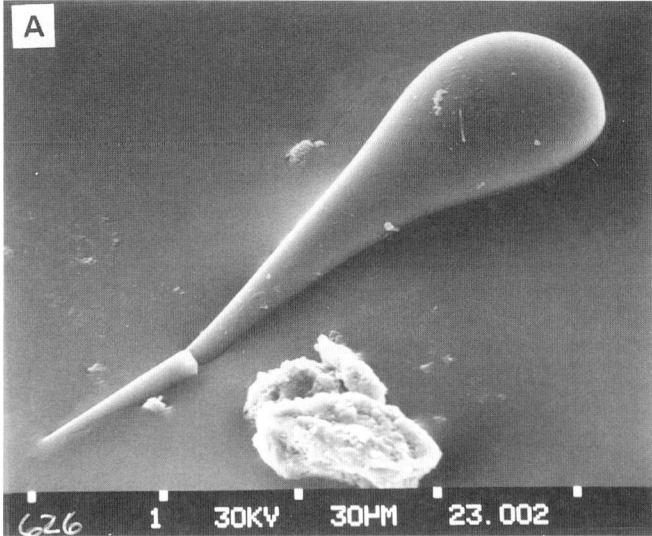
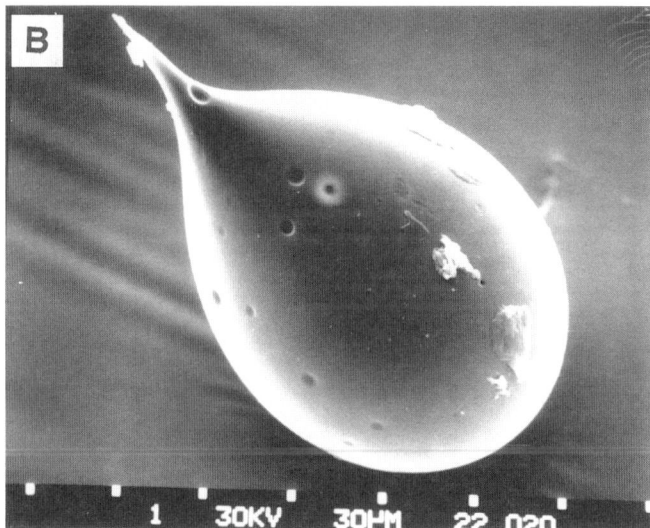
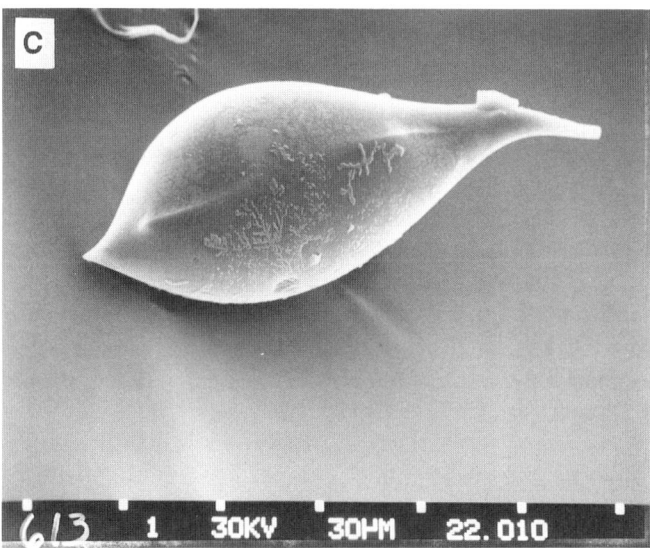

We have observed trees in growth position within a partially altered tuff in a Tertiary coal bed on the Kenai Peninsula, Alaska. Carbonized stumps ranging from a few centimeters to about 0.5 m in diameter are preserved in a 1-m-thick, partially altered ash layer within a coal bed (Fig. 27, A and B). The stumps occur several meters apart, and the finely laminated character of the ash suggests deposition either into shallow standing water or onto a dry surface. This same ash bed contains small logs (~10 cm in diameter) preserved as ash-filled casts, with only a thin carbonized rind to mark the original location of the outer surface.

Smaller (about 1 cm diameter) twigs are quite common in tonsteins, however; these are generally compressed parallel to the bedding, probably representing detrital plant material that accumulated during the ash fall. In the coal beds of the Ferron Sandstone Member (Triplehorn and Bohor, 1981), some of these twigs cut diagonally across the tonstein layers, possibly representing plants in growth position that have been deformed later after burial and compaction.

Leaves and other miscellaneous plant materials are often preserved in tonsteins. Leaves are most often found as impressions, but other plant fragments are either compressed or exhibit well-preserved plant tissues (Stach et al., 1982). Cell lumens may be filled with cryptocrystalline kaolinite or other minerals (Fig. 28, A, B, and C).

Reworking

Most tonsteins in coal beds were deposited as ash falls in standing water within a peat swamp. The water is presumed to have been fairly shallow (<0.5 m), or the swamp could not have supported continuous plant growth except as floating mats. Some tonsteins do show evidence of reworking in their upper portions, however, indicating enough depth of water to allow disturbance of the depositional surface, perhaps by storm runoff. An example of a partially reworked top in a thick tonstein was reported by Triplehorn and Bohor (1981) from the Ferron C coal bed in Utah.

ALTERATION OF VOLCANIC ASH TO TONSTEIN

Time of alteration

Most investigators consider the alteration of pristine volcanic ash to claystone (tonstein or bentonite) as a rather rapid geologic process. Srodon (1976) noted calcite concretions in

◂

Figure 26. SEM photos of aerodynamic glass droplets (kaolinized) showing delicate tails, similar to Pele's tears (Fisher and Schmincke, 1984, p. 102). Glades and Rock Springs #5 coal beds, Rock Springs Formation (Upper Cretaceous), Rock Springs, Wyoming.

Figure 27. Stumps in growth position penetrating a thick volcanic-ash layer. A: Several stumps washed free of ash by wave action; hammer for scale. B: Stump (coalified black feature) surrounded by thick volcanic ash (white). Outcrop photos along sea cliffs, Cook Inlet side of Kenai Peninsula, Alaska; Clamgulchian Stage type section (Miocene).

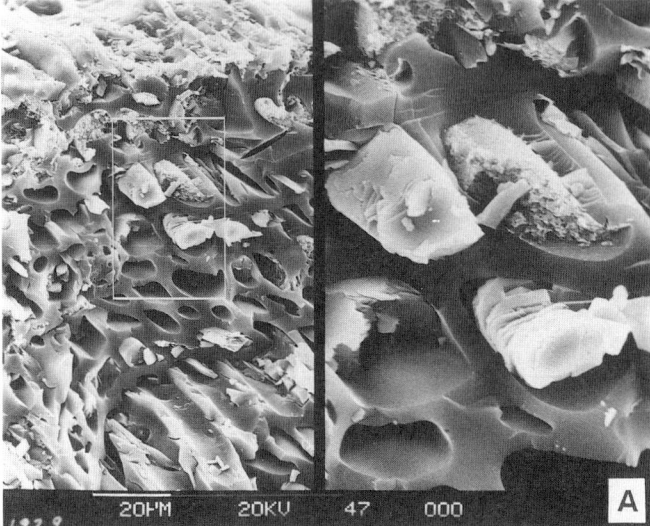
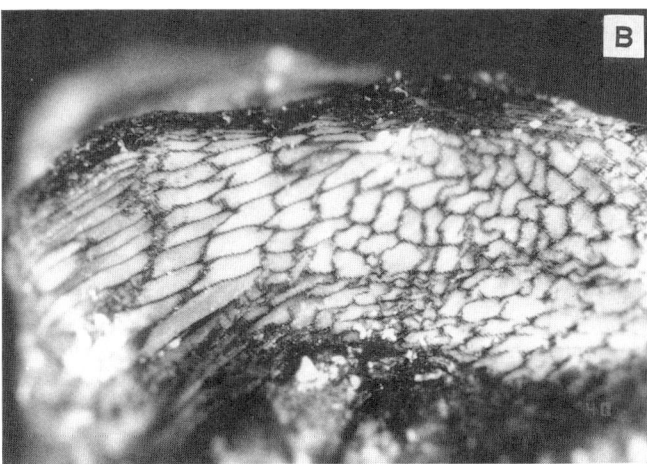

Figure 28. Microcrystalline kaolinite filling plant lumens. A: White microcrystalline kaolinite precipitated within dark organic plant lumens. Dual screen SEM photo; white box in left half is view enlarged in right half. Horsetooth Member, Dakota Group (Lower Cretaceous), Colorado. B: White kaolinite filling plant cells. Ordinary light photo; width of field = 2.3 mm. Flint clay in Olive Hill Clay (Lower Pennsylvanian), eastern Kentucky.

tonsteins from the Upper Silesian coal basin, Poland, that formed while the tuff material was still soft, prior to the formation of kaolinite. He concluded that clay-mineral formation probably occurred during early diagenesis, shortly after deposition of the ash. Burger (1966) observed brittle structural fractures in a Ruhr tonstein that must have occurred soon after its deposition, because immediately overlying units have normal stratification. He noted that the fracture (apparently caused by sliding of the unit) occurred after the sediment (ash) had been altered to claystone, because kaolinite crystals and vermicules cut by the fractures were already fully developed (as seen in thin section). Burger concluded from this example that clay-mineral formation in kaolin-coal tonsteins takes place very early during diagenesis.

Triplehorn and Bohor (1981) cited textural evidence from the "thick" tonstein of the Ferron C coal bed showing that alteration of ash to claystone took place prior to deposition of the overlying delta-front sandstone—a period estimated to have been a few thousand years. In contrast, glass and other minerals in some of the volcanic-ash beds in coals of Miocene age along the coast of the Kenai Peninsula, Alaska, have yet to alter completely to clay (Reinink-Smith, 1990b)—a period of time encompassing up to 15 m.y. Factors such as climate, hydraulic regimes in the depositional environments, and ash composition are clearly important contributors to the length of time needed to completely alter volcanic ash to a tonstein in coal-forming environments.

Influence of bed thickness

Climate, ash composition, fluid flow, and bed thickness are all factors in the diagenetic equation (Bohor, 1985). The effect of bed thickness has been studied in tonsteins of the Western Interior, particularly those of the Ferron Sandstone Member in Utah by Triplehorn and Bohor (1981); the following is taken from their work. The C coal bed of the Ferron contains several thin tonsteins (<10 cm thick) that are composed entirely of kaolinite in the clay-mineral fraction. However, this coal bed also contains a much thicker tonstein parting, ranging from 23 to 68 cm in thickness. All of the tonsteins in this coal bed are derived from rhyolitic tuffs having similar compositions. The thick tonstein (Fig. 29) displays a vertical zonation of clay minerals. The top and bottom few centimeters of this bed are completely altered to kaolinite, while the central portion consists of a mixture of discrete kaolinite and smectite. Zoning is also evident in the distribution of the primary volcanic minerals. In the top and bottom portions of this thick tonstein, plagioclase feldspars and biotite are absent and have presumably altered to kaolinite. However, these primary minerals are present in the central portion of the bed, along with kaolinite and smectite clay minerals. This mineralogical zonation appears to be a function of flushing (leaching), and not of primary deposition. Where the overlying delta-front sandstone cuts downward into the top of the thick tonstein, this claystone unit is composed only of kaolinite, quartz, and occasional sanidine grains (plus resistate trace minerals) from top to bottom. Smectite and unaltered plagioclase and biotite are not present anywhere in the thick tonstein at this outcrop location, presumably because of more effective flushing (leaching) and alteration due to close proximity of the overlying permeable sandstone.

Vertical variations in mineralogy of thick tonsteins can be ascribed to restrictions in water movement into and through the tuff during early diagenesis. Organic-rich, low-pH waters entering thick tuff beds from the compacting peat above and below quickly alter the bounding zones to kaolinitic clay; this "armoring" effect reduces permeability and hydraulic conductivities, limiting subsequent solutional exchange across the bed's boundaries. Leaching is thereby minimized in the central portion of these thick units, resulting in the retention of cations such as Mg^{++} and Ca^{++}. Smectite is stable under these conditions of autoalkalization, and begins forming in place of kaolinite as soon as leaching rates decrease and pH increases. Thin ash falls do not exhibit this vertical variation in mineralogy; they are converted to kaolinite throughout because the bounding zones converge before permeability can be substantially reduced and an "armoring" effect developed. The flushing (leaching) mechanisms operative in the development of

Figure 29. Thick tonstein in Ferron C coal bed (similar to cover photo). The thick tonstein is just beneath the massive sandstone. Ferron Sandstone Member, Mowry Shale (Cretaceous), Utah.

both thick and thin tonsteins are described diagramatically in Figure 30. This diagram seems to indicate lateral flow through the permeable ash beds, but in reality these solutions probably don't travel far laterally before rising to the swamp's surface along cracks, fractures, and permeable zones in the compacting peat.

Srodon (1976) found a similar mineralogical effect of thickness in tonsteins from the Silesian coal basin in Poland; i.e., greater kaolinite content at the top and bottom of thick beds, with only kaolinite throughout the thin beds. The role of flushing (leaching) was not stressed in Srodon's model, however. Forsman (1984) described thick rhyolitic ashes deposited in a marine environment that also display an "armoring" ef-

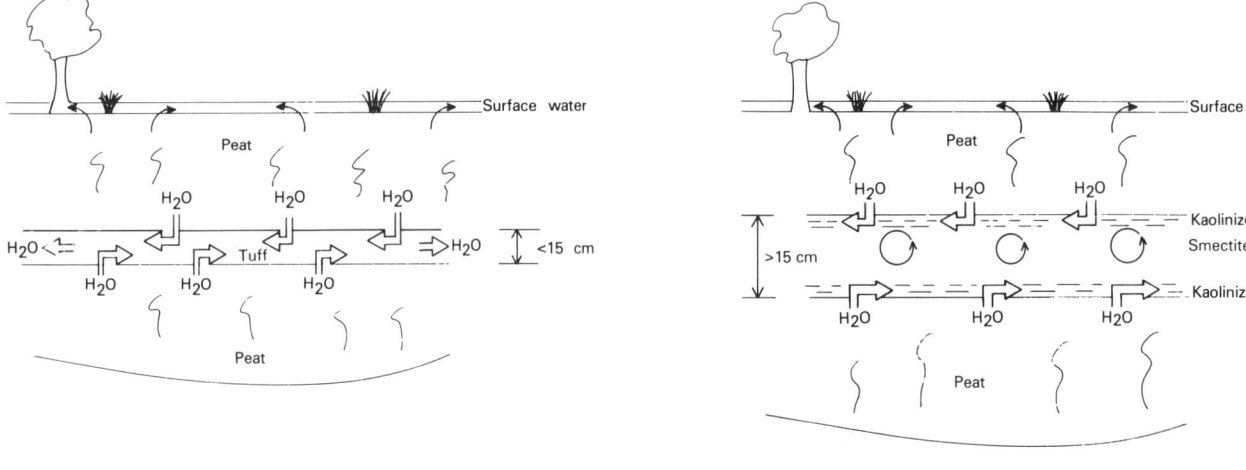

Figure 30. Effects of layer thickness on the alteration of volcanic tuffs to tonsteins. "Armoring" of top and bottom of thick tuffs by alteration of glass to kaolinite restricts further entrance (and exit) of fresh water and causes increased pH and autoalkalization of central portion (circular arrows) because of ionic buildup from solution of glass. This leads to the formation of smectite along with early-formed minor kaolinite in the interior zone.

fect, although here the armoring of the upper and lower contacts was due to smectite formation instead of kaolinite, and the central portion remained as unaltered glassy ash. He ascribed this situation to restriction of ion transport out of the enclosed ash because the bounding altered (smectitic clay) layers acted as semipermeable membranes.

Ash composition

We have discussed tonsteins whose progenitors were rhyolitic tuffs. However, tonsteins derived from more mafic tuffs do occur, and their mineralogy is often quite different from those formed from rhyolitic material. For example, Triplehorn et al. (1977) and Reinink-Smith (1990a) studied altered tuffs of intermediate composition in upper Tertiary coal beds on the Kenai Penninsula, Alaska. These tonstein partings often are smectitic, or are mixtures of smectite and kaolinite. Similar smectitic tonstein partings were described from Eocene coal beds in Washington by Reinink-Smith (1990a). These tonsteins are undoubtedly volcanic because they contain beta-quartz paramorph crystals, volcanic-glass shards and pumice, and euhedral crystals of zircon, apatite, amphibole and magnetite; no nonvolcanic components are present. The more mafic character of these tuffs is suggested by the presence of abundant hornblende and zoned plagioclase crystals, and by the absence of sanidine. All of the partings are less than 10 cm thick.

It is clear from their mineralogy that these Alaskan tuffs must have had a different source magma (dacitic) from that of the rhyolitic tuffs in coal beds of the United States Western Interior. Dacitic magmas are characteristic of island arcs, or of subduction zones at continental margins where basaltic oceanic plates are magmatized as they are subducted beneath the edge of the continental crust. Both island arcs and subduction zones are located off the coast of Alaska. These thin tuffs altered to smectite rather than kaolinite for either of two reasons (or both): (1) because of their mafic composition, the amount of alkaline earth cations was too great to be completely removed by leaching, and (2) the amount and rate of flushing was restricted, as indicated by the presence of unaltered glass shards and pumice in the tuff. Thus, original glass composition possibly can play a role in the diagenetic alteration of tuffs to tonsteins in coal beds under certain conditions. Francis (1961), however, observed basaltic tuffs in Scotland not directly associated with coal that have altered to kaolinite, showing that original glass composition is not the only factor controlling alteration products in tonsteins.

Alteration sequence

Most widespread tuffs are the product of explosive volcanism from rhyolitic magma sources. Therefore, let us consider a thin layer of distal rhyolitic ash deposited in a planar peat-forming swamp (we exclude raised bogs here) that was subsequently covered quickly with more peat, eventually resulting in the formation of a tonstein parting within a coal bed. The metastable fine-sized glass component will dissolve and precipitate as kaolinite if leached by large volumes of low-pH water containing humic and fulvic compounds from the peat. These thin, porous tephras initially act like sponges for water expelled from the compressing peat on both sides of the beds. The open system thus developed transports away excess ions and silica in solution, both laterally and vertically.

The chemical reaction paths involved in the alteration of volcanic ash to tonsteins under these conditions can be expressed conceptually in the following simplified forms, modified from Slaughter and Early (1965). These authors presented simple acid-base reaction equations with molar fractions of the reactants in their study (Slaughter and Early, 1965, p. 73–74) for those readers who wish to see this type of display, but our purpose here is to emphasize processes and not to complicate the story with chemical details.

$$\text{Rhyolitic glass} + H_2O \xrightarrow{\text{hydrolysis}} \text{hydrated aluminum silicate gel} + \text{cations (in solutions)}. \quad (1)$$

In an open, acidic environment with organic compounds present as a catalyst (Eberl and Hower, 1975), this intermediate hydrated aluminum silicate phase will go to kaolinite following the reaction path:

$$\text{Hydrated aluminum silicate (gel)} \xrightarrow{\text{organic acids}} \text{kaolinite (precipitated)} + \text{hydrated silica} + H_2O + \text{cations (solution)} \quad (2)$$

Some of the hydrated silica may react with hydrogen ions and precipitate quartz in the tuff bed; we have found examples of this secondary quartz in tonsteins (Fig. 31, A and B). Srodon (1976) reported finding secondary diagenetic quartz in altered tuffs.

In a partially restricted (low leaching) marine environment with an excess of Mg^{2+} from seawater (or a mafic ash), the intermediate hydrated aluminum silicate material will follow a reaction path to smectite:

$$\text{Hydrated aluminum silicate gel} + Mg^{2+} + Ca^{2+} \xrightarrow{\text{basic}} \text{smectite (precipitated)} + \text{hydrated silica (solution)} + H_2O. \quad (3)$$

If the system is closed and has a high salinity, zeolites will form:

$$\text{Hydrated aluminum silicate gel} + \text{alkaline cations} \xrightarrow{\text{highly basic}} \text{zeolite (precipitated)} + \text{hydrated silica (solution)} + H_2O. \quad (4)$$

We are postulating an intermediate hydrated aluminum silicate phase as the product of the initial hydrolysis reaction of volcanic glass (path 1). This intermediate product can then form either kaolinite or smectite, depending on the pH and

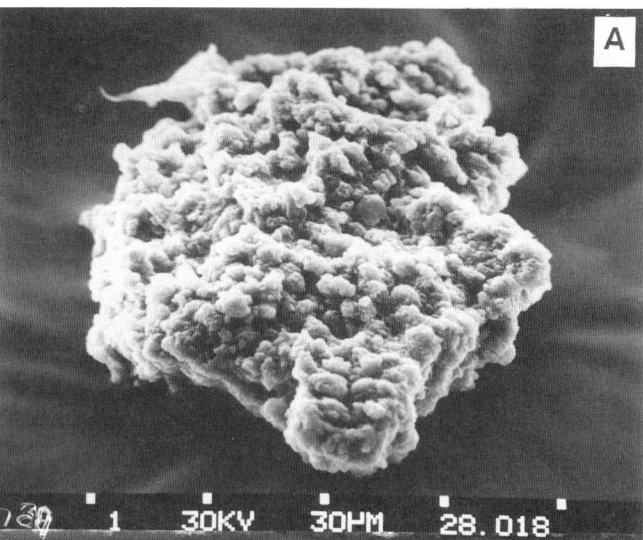

Figure 31. SEM photos of authigenic silica in tonsteins. A: Aggregated grain of authigenic silica. Ruffner flint clay (Middle Pennsylvanian), Greenbriar section, Charleston, West Virginia. B: Closeup of precipitated grains of silica forming aggregate showing rough anhedral surfaces of individual grains. Flint clay (Middle Pennsylvanian), Breathitt Formation, Mabie, Kentucky.

ionic content (freshness) of the water, the presence or absence of organic compounds, and the degree of leaching (open or closed system), which controls silica ($HSiO_4^+$) saturation and solubility. We have found small amounts of this unreacted hydrated aluminum silicate phase in some tonsteins (Fig. 32). This phase is neither kaolinite nor smectite and does not expand with glycol nor collapse with heating. Senkayi et al. (1984) proposed that in tonstein formation, volcanic glass alters first to smectite and then kaolinite, the latter presumably forming by diffusive reaction. This reaction path should lead

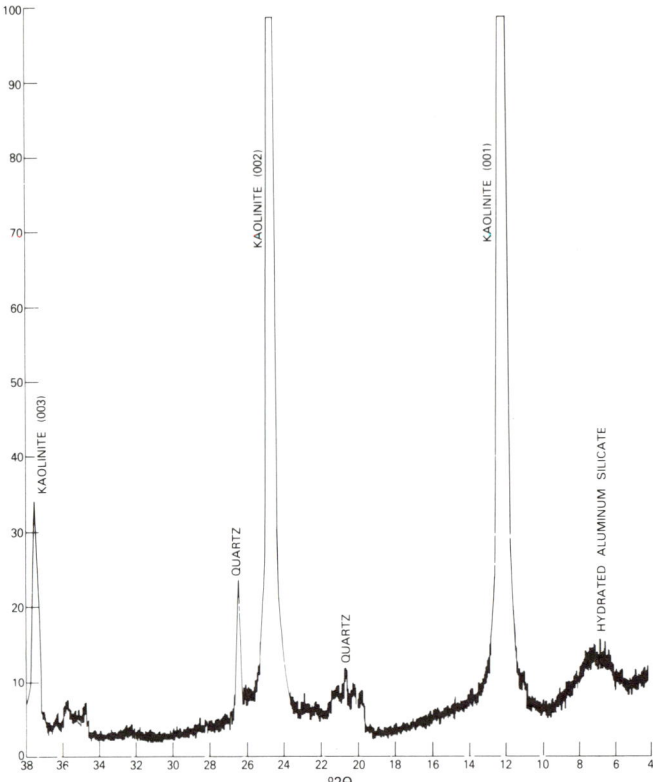

Figure 32. X-ray diffraction pattern (smear mount) of tonstein showing residual hydrated aluminum silicate intensity maximum (right) not affected by heat or glycol treatments.

to the production of an intermediate mixed-layer kaolinite-smectite phase; no such phase has been reported from tonsteins, however.

La Iglesia and Van Oosterwyck-Gastuche (1978) drew upon their own experimental studies and those of others to come to some interesting conclusions about the formation of kaolinite that seem to support the model outlined above in reaction-paths (1) and (2). These authors emphasized that kaolinite genesis involves a dynamic equilibrium related to the degree of leaching by extremely dilute solutions. The ionic concentration in the leachates depends more on the degree of leaching (flushing) than on the nature of the primary minerals, thus confirming the importance of the physicochemical conditions of weathering on clay-mineral genesis. La Iglesia and Van Oosterwyck-Gastuche (1978) further stated that the genesis of tonsteins is closely related to weathering by organic matter of the humic or fluvic type. Their studies also confirm the importance of a preexisting "dynamic" octahedral layer that orients the polymerization of the silica layer; this octahedral layer may be equivalent in part to the hydrated aluminum silicate component suggested by our reaction paths.

Hodder et al. (1990) suggested that the formation of clay minerals from modern volcanic tephra may follow a two-stage transformation, consisting first of diffusional glass hydration, followed by reactions forming clay minerals in rapidly weathering paleosol horizons. Other workers have established that first-order reaction kinetics predominate when these tephras are exposed to long periods of weathering. Note that both of these models were developed on subaerially weathered tephras (recent ash falls) in soil horizons, and may not be directly applicable to tephras altering in a peat-bog environment.

Only the finest vitric glass and labile primary mineral components seem to alter directly to clay minerals by solution-precipitation reactions. Diffusional stoichiometric replacement of larger, less-soluble glass clasts and nonlabile mineral grains by secondary minerals obviously occurs in tonsteins. Pseudomorphs of glass bubble junctions, splinters, and droplets by kaolinite and alumino-phosphate, and of feldspars and biotite by kaolinite, are evidence of this replacement reaction. However, kaolinitic pseudomorphs of glass fragments in tonsteins are rather rare, indicating the predominance of first-order solution-precipitation reactions in the development of these claystones. Diffusional processes as a whole become dominant in tonstein diagenesis only after burial to depths where ground water is no longer accessible and compaction has greatly reduced hydraulic conductivities into and within the bed.

In summary, rhyolitic volcanic ash alters to kaolinite when the depositional environment provides the following: (1) a high rate of leaching (open system); these dilute solutions maintain the low concentration of dissolved silica necessary for kaolinite formation at low temperature (La Iglesia and Van Oosterwyck-Gastuche, 1978); (2) organic matter of the humic or fulvic types, providing an acidic environment and organic catalysts (La Iglesia and Van Oosterwyck-Gastuche, 1978; Eberl and Hower, 1975); and (3) low Eh, pH, and ionic concentrations of solutions. These chemical conditions are usually met in a coal-forming environment; hence rhyolitic ashes usually alter to kaolinitic tonsteins there. However, smectitic tonsteins can develop where leaching is restricted (as in the central portion of thick ashes) or if the original ash composition is more mafic (high alkaline earth ion content). Shallow, organic-rich, brackish marine environments can produce kaolinitic bentonites from rhyolitic ash in an open system, in accordance with observations (Schultz, 1963; Pollastro and Martinez, 1985), even though the normal alteration products in low-leach-rate, alkaline, humic- and fulvic-poor marine environments are smectitic bentonites.

ENVIRONMENTAL EFFECTS OF THICK ASH FALLS

Air-fall deposition of voluminous quantities of volcanic ash into coal-forming environments can produce significant effects on the biological and hydrological regimes in these realms. The complete alteration of vitric volcanic ash to clay-rich tonsteins leads to a progressive reduction in thickness and porosity during alteration and compaction. Dokken (1987), using the deformation of fossil burrows, calculated an average compaction ratio of 3.86:1 in in an Ordovician bentonite. However, the maximum ratio shown by several of the samples in his study was 4.5:1, and this might be a more appropriate value for the compaction of volcanic ash deposited in seawater. Dopita and Kralik (1977) reported a compaction ratio of about 5:1 in a coal tonstein based on the reconstruction of a folded clastic sandstone dike in the seam. Thus, the original ash-fall layers in coal beds were probably almost five times thicker than the resulting tonsteins. Thick tonsteins, therefore, represent voluminous ash-fall deposits capable of overwhelming plant growth and modifying surface-water movement in peat swamps.

Water depths are typically shallow in peat-forming swamp environments. If the water were very deep, rooted plant growth could not be sustained and peat would not accumulate. Shallow water depths in peat swamps are substantiated by the lack of graded bedding in most tonsteins. Conversely, marine bentonites almost always show graded bedding because of the thick water column through which the ash must settle.

Plants, hydrology, and coal chemistry

The deposition of volcanic ash, particularly in a thick layer, influences plant growth and coal chemistry, as indicated by differences in coal petrography and geochemistry adjacent to tonsteins. Crowley et al. (1989) examined the petrology and geochemistry of coal immediately above and below the thick tonstein in the Ferron C coal in Utah (Triplehorn and Bohor, 1981) in order to determine the effect of thick ash falls on coal development. Maceral composition suggested well drained conditions in the peat swamp below the tonstein and poorly drained conditions above; the difference is presumably related to deposition of this thick ash. Leaching of the volcanic ash also led to changes in the chemical composition of the coal, which was enriched in a number of elements directly above or below the tonstein; this effect was also noted by Zielinski (1985) in tonsteins from the Powder River basin. Eble (1988) studied plant types immediately below and above a tonstein in the Hazard #4 coal of Kentucky. He found that floral differences above and below the tonstein reflect an interruption of plant growth, followed by a sequence of different plant types during the reestablishment of peat deposition.

Dufek (1987) showed that the voluminous ash precursor to the thick tonstein of the Ferron C coal bed in central Utah disrupted plant growth and succession to a much greater degree than did thinner ash falls in the same peat swamp. The original thickness of the air-fall ash layer forming the thick (up to 68 cm) tonstein is estimated to have been as much as 3.5 m. Palynologic analyses showed that the floristic community just prior to the ash fall was completely disrupted and replaced by a low-diversity, early successional community dominated by herbaceous ferns. Return to a plant community similar to that existing prior to the thick ash fall took a substantial period of time, represented by 65 cm of overlying coal (equivalent to as much as ~7 m of peat). The thick ash fall also caused flooding and ponding in certain areas of the swamp, resulting in detrital deposition that partially replaced peat accumulation in these areas. These pronounced floristic and hydrologic changes were noted only above the thick tonstein; three other thin tonsteins within the same coal bed did not cause any appreciable change in these two factors (Dufek, 1987).

Secondary minerals

Mineralogical changes can also be induced by thick ash-fall deposits. This is shown in an outcrop of the Frontier Formation on the Gray Cliffs along the Gros Ventre River, Wyoming (Love et al., 1948; Slaughter and Early, 1965). Here, an ~1.2 m white, siliceous claystone containing smectite, accretionary lapilli, and plant (fern) impressions displays a 30–40 cm red zone in its uppermost portion. This red color is caused by the presence of heulandite, a zeolite mineral related to clinoptilolite. It appears that the thick ash fall that formed this bed completely filled the swamp, disrupting the hydrologic regime by restricting surface drainage and producing ponding and evaporitic conditions. No peat accumulated in these ponds directly above this thick ash. These evaporative conditions and the resulting low leaching rates led to zeolite formation in the upper part of the ash bed and pervasive silicification of the rest of the bed within a closed system.

Diessel (1985) also noted that ash falls of several meters thick invariably caused prolonged interruptions to Permian peat accumulations in Australia and often contained secondary zeolites. Senkai et al. (1984) also observed zeolite (clinoptilolite) formation in the upper portion of a 20-cm-thick tonstein in an Eocene coal bed in Texas. They ascribed the zeolite formation to low leaching rates, citing Pevear et al. (1980), who observed that clinoptilolite formation in weathered bentonites was inversely related to degree of leaching.

STRATIGRAPHIC USES OF TONSTEINS

Air-fall volcanic-ash beds are extremely useful as correlation tools for several reasons. Individual ash layers were deposited in a geologic "instant" and thus form widespread isochron horizons of shorter duration than other event horizons, such as magnetic reversals and biological extinctions. Most rhyolitic-ash deposits contain phenocrysts of sanidine or biotite that can be dated by radiometric methods. Ash layers from large eruptions may extend over continent-sized areas and can be recognized thousands of kilometers from their source. Thus, volcanic-ash falls and the tonsteins formed from them are potentially useful for correlation both locally, as within an individual mine, and regionally between coal basins or on an intracontinental scale. Precise time control, relative or absolute (radiometric), can provide a framework for a variety of additional geologic studies, such as paleoenvironmental and paleogeographic reconstructions, chronostratigraphic calibration of fossil zones, and calculating rates of deposition.

Radiometric dating of primary minerals

Historical, fossils have provided the major relative age control in sedimentary rocks. Ages of coal beds and other nonmarine strata, however, are often not well determined because diagnostic marine fossils are absent and nonmarine fossils generally provide lower orders of stratigraphic precision. Where tonsteins are present, however, coal beds in nonmarine sections now may be dated radiometrically much more precisely than is possible using fossils (biostratigraphic zones). This radiometric precision is dependent on the sizes of the error bars, which increase in proportion to geologic age.

The principles for dating tonsteins are the same as for dating other volcanic materials, such as tephra or bentonite. A number of radiometric methods may be applied, including fission track, conventional K-Ar, Ar/Ar, Rb-Sr, and U-Pb (Fig. 33). Many potentially datable primary minerals are present in volcanic ash, but in tonsteins the most likely minerals to survive alteration are sanidine, plagioclase, zircon, and sometimes biotite, apatite, and hornblende. Whole-rock ages for tonsteins, such as the Rb-Sr ages reported by Bouroz et al. (1972), are less useful because tonsteins usually formed in open systems where elements can be introduced or removed by percolating solutions. U-Pb isotopic ages have been determined on zircon fractions isolated from bentonites (Samson et al., 1989) and this technique can also be applied to tonsteins. Individual zircon crystals can now be analyzed for U-Pb isotopic dating, greatly reducing the problems of inherited lead and contamination (Compston et al., 1992).

Conventional K-Ar methods have long been applied to bentonites (Folinsbee et al., 1961) and more recently to tonsteins (e.g., Damon and Teichmuller, 1971; Triplehorn et al., 1977, 1984; Turner et al., 1980, 1983; Marvin et al., 1986). However, the relatively low precision of this method often limits application of the data, and the inability to completely eliminate contamination and mineral alteration presents a serious problem. In addition, the relatively large sample required for analysis is usually impractical for some thin tonsteins or those recovered from cores.

A recent advance in radiometric dating, the $^{40}Ar/^{39}Ar$ technique, is a vast improvement over existing methods and provides perhaps the most exciting possibilities for tonstein studies. Analytical precision has been improved dramatically by the analysis of isotopic ratios rather than absolute amounts; step-wise heating permits evaluation of possible mineral alteration, and laser-fusion techniques permit precise age determinations from single, sand-sized mineral grains. Thus far, only a few results of $^{40}Ar/^{39}Ar$ dating of tonsteins are available. Lippolt and Hess (1985), Hess and Lippolt (1986), and Hess et al. (1988) have determined ages for sanidine and plagioclase in the Hazard #4 tonstein of Kentucky and in several tonsteins in Europe using the step-heating technique. Bohor et al. (1991) dated tonsteins from coal beds in the Dakota Formation (Cretaceous) of southern Utah using the $^{40}Ar/^{39}Ar$ continuous laser-fusion technique.

Calibration of fossil zones

Precise radiometric dating of tonsteins in coal-bearing strata can provide improved calibration of paleobotanical and palynological zones. This, in turn, could extend the usefulness of these fossil data for assigning approximate ages, even where volcanic-ash layers are not present. We recommend searching for datable tonsteins in stratigraphic sections containing important paleontological sequences as a means of cross calibration.

Conventional K-Ar and fission-track analyses of tonsteins have been related to palynology and paleobotany in southern Alaska and western Washington (Triplehorn et al., 1977, 1984; Turner et al., 1980, 1983). The utility of these studies would probably improve if the tonsteins were redated using $^{40}Ar/^{39}Ar$ techniques. An important application of the latter technique would be for determining the age relation of Carboniferous fossil floras of Europe and eastern North America.

Isochron horizons

Tonsteins are isochronous stratigraphic units. Preserved beneath each of them is a surface that existed at a moment in time. The overlying sediment similarly reflects conditions shortly after the ash fall, because preservation of the ash layer requires almost immediate burial. In peat beds, this takes place by resumption of plant growth, probably within a year or so. Rapid deposition and immediate burial requires that strata above and below a given tonstein are essentially the same age as the tonstein. Therefore, lateral variations in the underlying coal reflect differences in topography and depositional environments of the peat swamp, whereas organic petrology of the overlying peat reflects sequential stages in the reestablishment of plant growth.

Coals are commonly sampled with the assumption that laterally separated vertical sections through the same bed represent similar time intervals. When traced laterally, however, tonsteins sometimes extend diagonally through a coal bed, indicating that this assumption may not always be valid. In the Ferron C coal of Utah (Ryer et al., 1980), the identification of four laterally continuous tonsteins permitted the coal bed to be divided into five isochronous subunits (Fig. 34). Mapping of these isochronous units showed that the position of maximum peat accumulation shifted laterally with time, and that the base of the coal bed is time transgressive. Such data have important geologic and economic implications. In the absence of time control, such as that established with tonsteins, lateral variations in geochemistry, mineral matter, paleobotany, or coal petrography may be interpreted as geographical variations in the

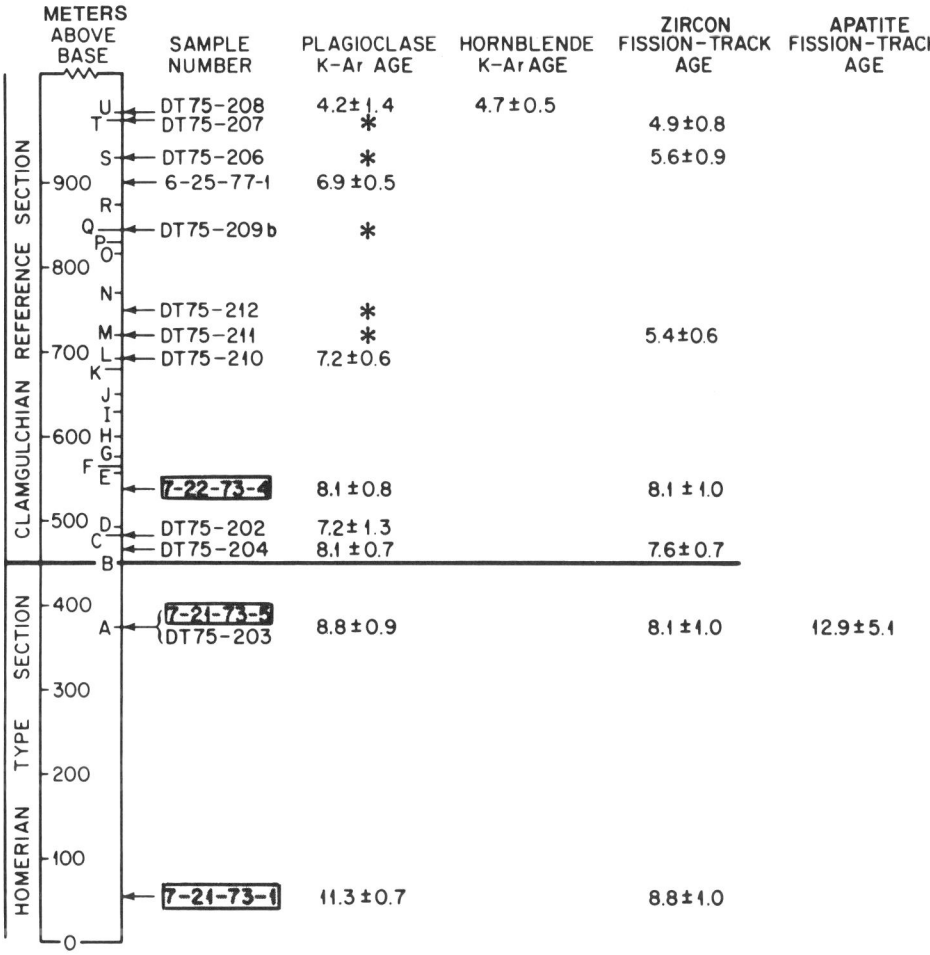

Figure 33. Radiometric dating of Tertiary (Neogene) coals, Kenai Peninsula, Alaska, by K-Ar and fission-track methods. From Triplehorn et al. (1977).

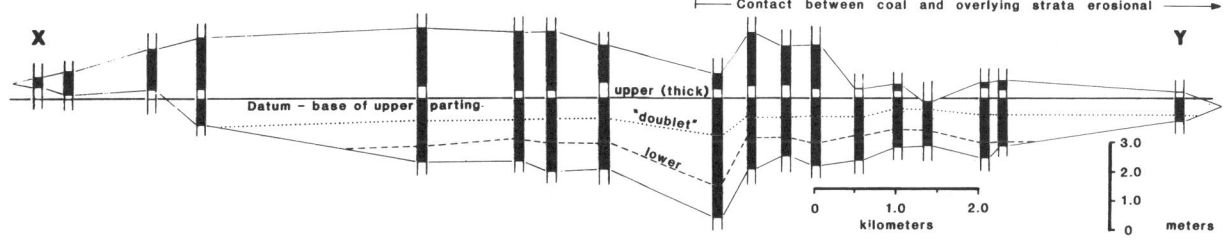

Figure 34. Cross section of Ferron C coal bed (Upper Cretaceous) in central Utah. Datum is base of thick tonstein. Seaward is to the right, landward is to the left. From Ryer et al. (1980).

peat accumulation environment, rather than as undetected age differences between sampling horizons.

Correlation of individual units

The use of tonsteins as marker beds requires that any given tonstein can be recognized in widely separated localities, and that it is distinguishable from other closely associated tonsteins. In some cases, there are obvious differences between individual tonsteins; in other cases, a series of tonsteins may appear to be very similar. The following criteria can be applied to distinguish individual tonsteins; this list is illustrative, not exhaustive.

Lithologic succession. The relation of tonsteins to distinctive adjacent beds is fundamental, but may not carry for long distances, and depends on areal distribution and preservation of the ash fall. Interruptions in ash-bed continuity and changes in bounding lithologies are to be expected, but this should not hinder identification if the changes are not abrupt and exposures are good (see Burger, 1971, for an example of tonstein correlations in Ruhr coalfields using this technique). Closely coupled tonstein sequences such as doublets are often useful in correlation, especially if substantially separated vertically from other tonsteins in the coal bed.

Field appearance (e.g., thickness, color, and texture). Visual appearance is generally reliable, but there are exceptions. For example, most tonsteins are white–weathering on outcrop, but some are not; weathered appearance is not helpful when studying cores or fresh exposures. Thickness can change over long distances, but is usually fairly constant within a coal field. Texture may also change laterally, as shown by Burger (1971) from coal fields of the Ruhr Valley.

Bulk clay mineralogy. The great majority of thin tonsteins that were derived from a rhyolitic precursor are composed of well-crystallized kaolinite, and adjacent tonsteins (eruptive events) tend to have similar clay compositions. Occasionally tonsteins are found that have clay compositions different from those of their neighbors. These disparities are probably due more often to postdepositional factors that affect the water-rock interactions of these particular beds, such as a relatively impermeable peat enclosing the tonstein layer, rather than to abrupt changes of erupted-ash composition.

Bulk and trace-element chemistry. Determination of the major element bulk chemistry of tonsteins usually has little application, because such material has been lost and some new material added during alteration and later diagenesis. However, Spears and Kanaris-Sotiriou (1975) analyzed for the (relatively) immobile elements Ti and Al in tonsteins to determine the composition of the original ash, and thus the type of source magma. In addition, studies of the rare earth elements (REE) in tonsteins (Bohor and Meier, 1990; Kendrick, 1985) are useful for determination of volcanic origin and weathering schemes. Figure 35 shows REE patterns of some tonsteins from the Western Interior of North America, as measured by induction-

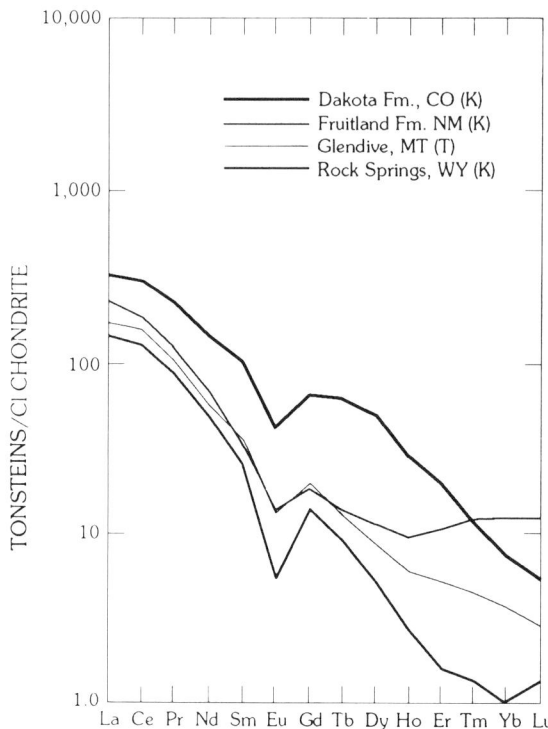

Figure 35. REE patterns of four tonsteins from the Western Interior of North America analyzed by ICP-MS. Note the negative europium anomaly characteristic of silicic volcanism. From Bohor and Meier (1990).

coupled plasma mass spectrometry (ICP-MS). Trace element chemistry of bulk samples (Kolata et al., 1986) and individual mineral species (Samson et al., 1988) has also been used to distinguish individual marine Ordovician K-bentonite beds in the mid-continent region.

Relative abundance of nonclay minerals. Mineral species can be very diagnostic, but only primary volcanic minerals should be employed. Total quartz content is easily determined (by XRD), and quartz is generally resistant to solution or alteration under acidic conditions. Normally, the quartz content of tonsteins is rather low compared to shales, and an unusually large amount of quartz indicates that the bed is probably reworked, or at least partially contaminated by detrital sediment. A simple field test for a tonstein consists of chewing a piece between the teeth. Tonsteins feel smooth to the teeth; shales and other detrital partings will feel gritty in this test due to their greater content of quartz and other nonclay minerals.

The presence of unusually large amounts of beta-quartz paramorph crystals or idiomorphic zircons can be useful for discrimination between tonsteins. Rare or trace minerals, such as topaz, garnet, monazite, or allanite, can also be use to discriminate between individual tonsteins (Bohor et al., 1979).

Elemental composition of mineral species. Lerbekmo et al. (1975) used the major element composition of ilmenite

(Fe, Ti, Mg) to successfully distinguish two Holocene ash lobes that were otherwise identical in mineralogy. However, ilmenite in tonsteins may not be as useful, because it may alter to anatase or rutile over time.

Samson et al. (1988) determined REE concentrations in primary apatite phenocrysts by instrumental neutron-activation analyses and used these data to correlate Ordovician K-bentonites over long distances in the upper Mississippi Valley. Sampson et al. emphasized that analysis of primary phenocrysts rather than bulk samples for correlation eliminates the contamination problem often associated with bulk samples. The electron microprobe also has proven useful for trace elemental studies of inclusions within minerals. Belkin and Rice (1989) successfully analyzed silicate-melt glass inclusions within various primary volcanic minerals from a tonstein with the microprobe in order to determine the composition of the parent magma. This technique also seems promising as a means of "fingerprinting" individual tonsteins.

Morphology of mineral species. Kowallis and Christiansen (1989) and Kowallis et al. (1989) used the morphology of zircons, based on Pupin's (1980) classification scheme, to distinguish between a variety of Mesozoic and Tertiary altered ash-fall layers (bentonites): in addition, they used Pupin's work on the shapes of zircons in igneous rocks to determine composition and pressure-temperature conditions in the parent magmas of these ashes.

Recording eruptive histories

Thick coal beds may contain numerous thin tonstein layers representing multiple eruptive pulses of the source volcano. Examples that we have studied include the Late Cretaceous "Big Dirty" coal bed near Livingston, Montana (7 m of coal containing at lest 34 tonstein partings); another "Big Dirty" (many thick coals rich in partings are called "Big Dirty" by miners) coal bed of the Paleocene Fort Union Formation in the Bull Mountains near Roundup, Montana, that contains 15 tonstein partings in <3 m of coal (Fig. 36A and back cover photo); and a possibly correlative Paleocene coal bed (designated the "W" bed by stratigraphers) containing 15 partings in ~3 m of coal in the Hell Creek area of Montana.

A peat swamp that continues to accumulate organic material over a relatively long period of time performs like a strip-chart recorder for volcanic events during its lifetime. If the timing is right, one thick coal bed can record the entire history of one or more eruptive events upwind of the swamp, clearly delineating the individual eruptions in chronologic order. Time intervals between these eruptive events can be estimated by measuring coal thickness between tonsteins. Changes in the erupted magma composition as a function of time can be studied in the variation of primary minerals and their inclusions within the multiple tonstein layers. Marine units can also record multiple ash falls (Fig. 36B), but in this environment contamination by detrital material, reworking and bioturbation, and coa-

Figure 36. Multiple volcanic eruptions recorded by tonsteins and bentonites. A: 15 crystal tonsteins (thin white streaks) in 3.3-m-thick "Big Dirty" coal bed, Fort Union Formation (Paleocene), Roundup, Montana. Vertical white scale on outcrop is a 6–in. ruler. (Also see back cover photo). B: Multiple bentonites (white layers) in road cut of marine Mowry Shale (Upper Cretaceous) near Lander, Wyoming.

lescence of multiple eruptive units with graded bedding inhibit the application of this technique (Reinink-Smith, 1990b). These problems are not usually encountered in tonsteins, making them a better choice for this type of study.

The most likely place to find thin, distal, volcanic-ash layers preserved is in coal beds. In most marine environments, biological activity and erosive currents combine to obliterate thin volcanic-ash layers; thus, they are not likely to be preserved here as discrete units (a point made in many studies; e.g., Ruddiman and Glover, 1972). In contrast, most peat-forming environments have so little physical and biological activity that extremely thin (~1 mm) ash layers are preserved intact. Incorporation within peat isolates the volcanic material from terrigenous water-laid sediment, making the resulting tonstein visibly distinctive and relatively uncontaminated. In addition, the rapid regrowth and accumulation of vegetation allows for the preservation of separate eruptive deposits as distinct strata (Borchardt et al., 1973).

Intercontinental correlations

The Plinian eruptions of silicic volcanoes can expell significant volumes of ash into the upper atmosphere, and fine aerosol particles may be carried around the world several times (Lamb, 1970). Langway et al. (1988) reported a volcanic acid aerosol signature from an A.D. 1259 eruption preserved in ice cores in both Greenland and Antarctica. Although it does not involve particulate matter, this study indicates that high-altitude winds are capable of distributing volcanic material to high latitudes in both hemispheres. It seems likely that the largest volcanic eruptions could have created nearly world-wide aerosol veils and probably affected the global climate (Porter, 1981; Newell and Walker, 1981), although the intensity and duration of such effects are unknown. Eruptive events that have spread ash deposits over many hundreds or even thousands of kilometers have been documented. Therefore, tonsteins should be correlatable over these distances, and this was demonstrated stratigraphically in western Europe (Fig. 37) by Burger and Damberger (1985).

Samson et al. (1989) analyzed a zircon fraction from an Ordovician K-bentonite from the Mississippi Valley by the U-Pb isotopic method and showed that the concordant date obtained was identical within analytical error with those derived from Ordovician K-bentonites from Sweden, suggesting a common volcanic source. Huff et al. (1992) have shown a correlation between two of these Ordovician K-bentonites, the Millbrig in eastern North America and the "Big Bentonite" in Baltoscandia (Sweden), based on discriminant function analysis of trace element data. According to Huff et al. (1992), the correlation of these two bentonites between North America and northern Europe suggests an eruptive source in an intermediate position that deposited a meters-thick ash layer over an area of several million square kilometers. This ultra-Plinian eruptive event is thus one of the largest ever recorded. Most tonsteins originated from eruptions much smaller than this Ordovician event, but the intercontinental proportions of the ultra-Plinian eruption suggests that correlation of certain tonsteins between eastern North America and western Europe may be feasible.

SUMMARY AND CONCLUSIONS

We define tonsteins as altered, distal, air-fall volcanic-ash layers commonly preserved in coal-accumulating environments in nonmarine strata. The clay mineralogy of these thin claystone layers is usually kaolinitic, with phenocrysts of primary volcanic minerals floating in a clay matrix. The authigenic nature of the kaolinite is demonstrated by its purity and high degree of crystallinity, its occurrence as fragile vermicules, and its pseudomorphing of volcanic-glass fragments and labile mineral grains. The phenocrystic component usually comprises a typical silicic volcanic suite, including beta-form quartz paramorphs, sanidine, and euhedral zircon, biotite, and (sometimes) apatite crystals. Analysis of these phenocrysts is essential for the identification and characterization of tonsteins, and the geologic uses of these layers is predicated on the basis of their origin from air-fall volcanic ash. Therefore, preparatory methods designed to separate these phenocrysts from the clay matrix should be an integral part of any tonstein study. A processing scheme that uses mechanical, ultrasonic, and chemical dispersion techniques to suspend the clay-mineral fraction for removal by elutriation has been successful in these investigations. Individual primary volcanic phenocrysts thus isolated can then be studied optically and with a scanning electron microscope.

The composition and texture of tonsteins are functions of original ash composition, bed thickness, geochemical conditions in the depositional environment, hydraulic conductivities (rate of leaching), and the course of early and late diagenesis. Diagenesis not only controls the alteration products of volcanic glass and mineral phases, but also the type and amount of secondary minerals found in tonsteins. Thin rhyolitic ash layers deposited in peat swamps typically alter to kaolinitic tonsteins. This environment provides humic and fulvic matter, low pH and Eh, and high leach rates with dilute (fresh) solutions. These factors provide optimum conditions for solution of silicic glass and direct precipitation of kaolin minerals, which are stable under these conditions at low temperatures. However, nearshore marine environments can sometimes provide similar conditions, and rhyolitic ash deposited here may alter to kaolinitic bentonite, rather than the usual smectitic variety.

Abundant evidence of rapid conversion of volcanic glass to clay in tonsteins, clay "armoring" of thick ash beds, no evidence of mixed layers of kaolinite-smectite, and a general lack of preserved pyroclastic textures in mature tonsteins all indicate that leaching and solution and/or precipitation are the dominant processes in the initial conversion of volcanic tuff into claystone. These first-order reactions predominate in the

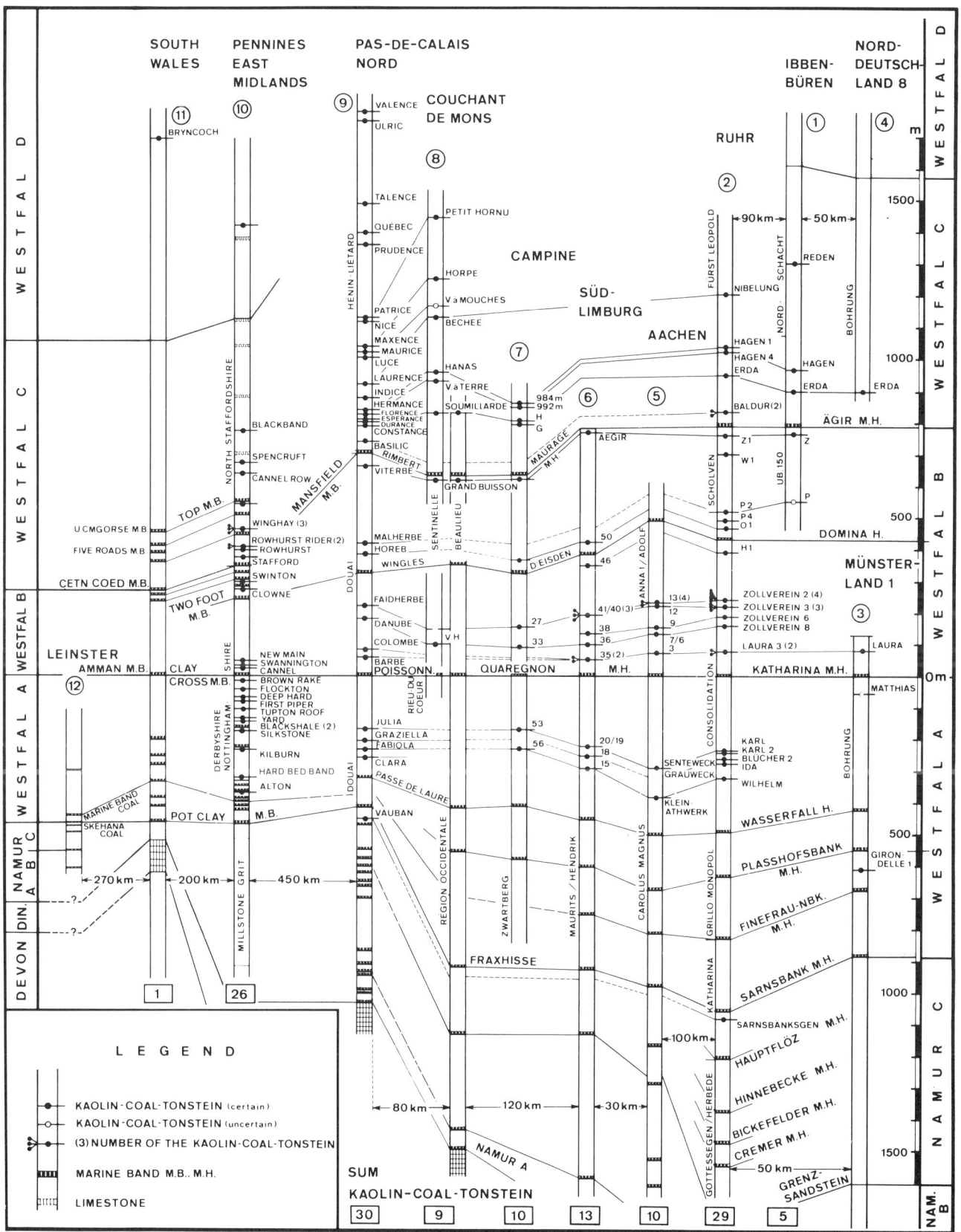

Figure 37. Correlation of Pennsylvanian strata from northern Germany to Wales using tonsteins. From Burger and Damberger (1985).

early conversion of glass to clay, whereas diffusional processes may become dominant later on as hydraulic conductivities decrease upon deeper burial.

Once tonstein partings are identified as altered volcanic-ash falls by field expression and mineralogy, they can be used as isochrons for stratigraphic correlation, bed zoning, and radiometric dating, much as their marine analogs (bentonites) have been used for years. However, tonsteins may be even more useful to geologists than marine bentonites, because tonsteins occur in poorly fossiliferous nonmarine strata, where isochrons and extensive marker beds are rare. Calibration of biostratigraphic zones by radiometric dating of tonsteins is one important application of these beds that should be pursued; "fingerprinting" individual tonstein layers for correlation purposes is another. Utilization of the full potential of tonsteins has just begun, and the future holds much promise for their application to the many and varied problems in nonmarine rocks that require isochrons and datable marker beds for their solution.

ACKNOWLEDGMENTS

We thank R. Pollastro for his thoughtful reviews, which contributed greatly to the quality of the finished product. We also thank the other reviewers, G. Ludvigson, B. Witzke, and D. A. Spears, for their efforts, which resulted in improvement of the manuscript. We specifically knowledge the field and laboratory assistance of R. Pollastro, R. Phillips, C. Sanderson, W. Stang, D. Schultz, and W. Betterton, and the contributions over the years of various other colleagues at the U.S. Geological Survey in Denver, Colorado, and in Reston, Virginia.

REFERENCES CITED

Addison, R., Harrison, R. K., and Land, D. H., and Young, B. R., 1983, Volcanic tonsteins from Tertiary coal measures, east Kalimantan, Indonesia: International Journal of Coal Geology, v. 3, p. 1–30.

Banfield, J. F, and Eggleton, R. A., 1990 Analytical transmission electron microscope studies of plagioclase, muscovite, and K-feldspar weathering: Clays and Clay Minerals, v. 38, p. 77–89.

Bateson, J. H., 1965, Accretionary lapilli in geosynclinal environment: Geological Magazine, v. 102, p. 1–7.

Belkin, H. E., and Rice, C. L., 1989, A rhyolite ash origin for the Hazard No. 4 flint clay (Appalachian basin): Evidence from silicate melt inclusions: Geological Society of America Abstracts with Programs, v. 21, no. 6, p. A360.

Bischof, G., 1863, Lehrbuch der chemischen und physikalischen Geologie (second edition): Bonn, Germany, A. Marcus.

Blatt, H., Middleton, G., and Murray, R., 1980, Origin of sedimentary rocks (second edition): Englewood Cliffs, New Jersey, Prentice-Hall, Inc., 782 p.

Bohor, B. F., 1983, Clay mineralogy of a Cretaceous-Tertiary boundary claystone from Montana, in Abstracts and Programs, 32nd Annual Clay Minerals Conference: Buffalo, New York, Clay Minerals Society, p. 48.

Bohor, B. F., 1985, Diagenesis of rhyolitic ash in coal-forming environments [abs.], in Book of Abstracts, International Clay Conference, Denver, Colorado: Boulder, Clay Minerals Society, p. 27.

Bohor, B. F., and Meier, A. L., 1990, REE abundances of tonsteins and K-T boundary claystones by ICP-MS [extended abs.], in Lunar and Planetary Science XXI: Houston, Texas, Lunar and Planetary Institute, p. 109–110.

Bohor, B. F, and Pillmore, C. L., 1976, Tonstein occurrences in the Raton coalfield, Colfax County, New Mexico, in New Mexico Geological Society, 27th Field Conference, Guidebook: Soccoro, New Mexico Bureau of Mines and Mineral Resources, p. 177–183.

Bohor, B. F., and Triplehorn, D. M., 1981, Volcanic origin of the flint clay parting in the Hazard #4 (Fire Clay) coal bed of the Breathitt Formation in eastern Kentucky, in Coal and coal-bearing rocks of eastern Kentucky (Geological Society of America Annual Meeting, Coal Division field trip guidebook): Lexington, Kentucky Geological Survey, p. 49–54.

Bohor, B. F., and Triplehorn, D. M., 1984, Accretionary lapilli in altered tuffs associated with coal beds: Journal of Sedimentary Petrology, v. 54, p. 11–25.

Bohor, B. F., Hatch, J. R., and Hill, D. J., 1976, Altered volcanic ash partings as stratigraphic marker beds in coals of the Rocky Mountain region: American Association of Petroleum Geologists Bulletin, v. 60, p. 651.

Bohor, B. F., Phillips, R. E., and Pollastro, R. M., 1979, Altered volcanic ash partings in Wasatch Formation coal beds of the northern Powder River basin: Composition and geologic applications: U.S. Geological Survey Open-File Report 79-1203, 21 p.

Bohor, B. F., Dalrymple, G. B., Triplehorn, D., and Kirschbaum, M., 1991, Argon/Argon dating of tonsteins from the Dakota Formation, Utah [abs.]: Geological Society of America Abstracts with Programs, v. 25, no. 5, p. A85.

Borchardt, G. A. Norgren, J. A., and Harward, M. E., 1973, Correlation of ash layers in peat bogs of eastern Oregon: Geological Society of America Bulletin, v. 84, p. 3101–3108.

Bouroz, A., 1962, Sur la pluralite d'origine des tonsteins: Société Géologique du Nord Annales, v. 82, p. 77–94.

Bouroz, A., Chalard, J., and Dolle, P., 1953, Extension geographique et valeur stratigraphique des tonsteins du bassin houiller du Nord de la France: Société Géologique du Nord Annales, v. 73, p. 98–141.

Bouroz, A., Roques, M., and Vialette, Y., 1972, Etude de la cinerite au sommet de la zone 2 du bassin des Cevennes, in Paris, Bureau de Researches Geologie et Minieres Memoir 77, p. 503–507.

Bouroz, A., Spears, D. A., and Arby, F., 1983, Review of the formation and evolution of petrographic markers in coal basins: Société Géologique du Nord Memoires, Tome XVI, 115 p.

Burger, K., 1955, Ergebnis der petrographischen Untersuchung eines Tonsteinfundes aus dem Floznebengestein des Ruhrkarbons [Result of petrographic investigation of a tonstein find from the country rock of the Ruhr Carboniferous]: Gluckauf, v. 91, p. 988–990.

Burger, K., 1958, Uber 2 neue Tonsteinfunde aus dem Floznebengestein des Ruhrkarbons: Bergfreiheit, v. 23, p. 49–51.

Burger, K., 1964, Stratigraphie und Petrographie des neuen Kaolin-Kohlentonstein des Flores H (EB) in den Oberen essener Schlichten (Westfal B) des Ruhrkarbons [Stratigraphy and petrology of the new kaolin-coal-tonstein from the H1 (EB) seam in the upper Essen beds (Westphalian B) of the Ruhr Carboniferous]: Fortschritte in der Geologie von Rheinland und Westfalen, v. 12, p. 451–472.

Burger, K., 1966, Zur entstehung der kaolinit-formentypen (graupen und kristalle) in kaolin-kohlen-tonsteinen: Geologische Mitteilungen, v. 6 (Breddin Festschriff), p. 43–86.

Burger, K., 1971, Monographie des Kaolin-Kohlentonsteins Zollverein 8 in den Essener Schichten (Westfal B1) des Niederrheinisch- Westfalischen Steinkohlenreviers; Teil I: Der Kaolin-Kohlentonstein Zollverein 8, sein Auftreten im Floz- und Schichtenverband, seine makroskopische Ausbildungsform und sein Bedeutung fur die Stratigraphie im Ruhrkarbon: Forschungsberichte des Landes Nordrhein,-Westfalen: Cologne-Opladen,

Westdeutscher Verlag, no. 2125, 95 p.

Burger, K., 1979, Vorkommen und stratigraphische verteilung des Kaolin-Kohlentonstein in der kohlen-revieren der erde: Congres International de Stratigraphie et Geologie du Carbonifere, 8th, Moscow, 1975, Comptes Rendus, v. 5, p. 21–31.

Burger, K., 1985a, Petrography and chemistry of tonsteins of the coal basins of Western Europe and North America: Congres International de Stratigraphie et de Geologie du Carbonifere, 9th, Champaign-Urbana, Illinois, 1979, Comptes Rendus, v. 4, p. 449–466.

Burger, K., 1985b, Die Kohlentonsteine Im Niederrheinisch-Westfalischen Steinkohlenrevier. Erkenntnisstand 1983: Congres International de Stratigraphie et Geologie du Carbonifere, 10th, Madrid, 1983, Comptes Rendus, v. 4, p. 211–234.

Burger, K., 1990, Vulkanogene Glasscherben-Relikte in Kohlentonsteinen des Saar-Lothringer Oberkarbons sowie Herkunft und Menge der Pyroklastika: Geologische Rundschau, v. 79, p. 659–691.

Burger, K., and Damberger, H. H., 1985, Tonsteins in the coal fields of Western Europe and North America: Congres International de Stratigraphie et Geologie du Carbonifere, 9th, Champaign-Urbana, Illinois, 1979, Comptes Rendus, v. 4, p. 433–448.

Burger, K., and Stadler, G., 1984, Vulkanogene Glasscherbenrelikte im Z-1-Kohlentonstein des Ruhrkarbons [Volcanogenic glass splinter remnants in the Z-1 coal tonstein of the Ruhr Carboniferous]: Fortschritte in der Geologie von Rheinland und Westfalen, v. 32, p. 171–186.

Burger, K., and Wolf, M., 1987, Remains of volcanic glass splinters in the tonsteins of the Upper Carboniferous in the Saar-Lorraine Basin: Congres International de Stratigraphie et Geologie du Carbonifere, 11th, Beijing, 1987, Comptes Rendus, section 5, p. 218–219.

Burger, K., Zhou, Y., and Tang, D., 1990, Synsedimentary volcanic-ash-derived illite tonsteins in Late Permian coal-bearing formations of southwestern China: International Journal of Coal Geology, v. 15, p. 341–356.

Calvert, C. S., 1984, Simplified, complete CsCl-hydrazine-dimethylsulfoxide intercalation of kaolinite: Clays and Clay Minerals, v. 32, p. 125–130.

Chamley, H., 1989, Clay sedimentology: New York, Springer-Verlag, 623 p.

Clocchiatti, R., 1975, Les Inclusions Vitreuses des Cristaux de Quartz: Paris, Société Géologique de France, Memoire 122, nouvelle serie, Tome LIV, 96 p.

Compston, W., Williams, I. S., Kirschvink, J. L., Zichao, Z., and Guogan, M., 1992, Zircon U-Pb ages for the Early Cambrian time scale: Geological Society of London Journal, v. 149, p. 171–184.

Crowley, S. S., Stanton, R. W., and Ryer, T. A., 1989, The effects of volcanic ash on the maceral and chemical composition of the C coal bed, Emery Coal Field, Utah: Organic Geochemistry, v. 14, p. 315–331.

Damon, C., and Teichmuller, R., 1971, Das absolute Alter des sanidinfuhrenden kaolinishcen Tonsteins im Floz Hagen 2 des Westfal C im Ruhrrevier: Fortschritte in der Geologie von Rheinland und Westfalen, v. 18, p. 53–56.

Diessel, C.F.K., 1963, On the petrography of some Australian tonsteins: Clausthal-Zellerfeld, Max Richter Festschrift, p. 149–166.

Diessel, C.F.K., 1985, Tuffs and tonsteins in the coal measures of New South Wales, Australia: Congres International de Stratigraphie et Geologie du Carbonifere, 10th, Madrid, 1983, Comptes Rendus, v. 4, p. 197–210.

Dimanche, F., and Bartholome, P., 1976, The alteration of ilmenite in sediments: Mineral Science and Engineering, v. 8, p. 187–201.

Dokken, K., 1987, Trace fossils from Middle Ordovician Platteville Formation, in Sloan, R. E., ed., Middle and Late Ordovician lithostratigraphy and biostratigraphy of the Upper Mississippi Valley: St. Paul, Minnesota Geological Survey Report of Investigations 35, p. 191–196.

Donaldson, C. H., and Henderson, C.M.B., 1988, A new interpretation of round embayments in quartz crystals: Mineralogical Magazine, v. 52, p. 27–33.

Dopita, M., and Kralik, J., 1977, Coal tonsteins in Ostrava-Karvina coal basin (Uhelne tonsteiny Ostravsko-karvinskeho reviru): Ostrava, Czechoslovakia, 213 p.

Dufek, D. A., 1987, Palynological analysis of floral changes caused by repeated volcanic ash burial of a coal-forming Upper Cretaceous peat swamp, Utah, U.S.A. [Ph.D. thesis]: East Lansing, Michigan State University, 232 p.

Eberl, D. D., and Hower, J., 1975, Kaolinite synthesis: The role of the Si/Al and (alkali)/(H+) ratio in hydrothermal systems: Clays and Clay Minerals, v. 23, p. 301–309.

Eberl, D. D., Srodon, J., Kralik, M., Taylor, B. E., and Peterman, Z. E., 1990, Ostwald ripening of clays and metamorphic minerals: Science, v. 248, p. 474–475.

Eble, C. F., 1988, Palynology and paleoecology of a Middle Pennsylvanian coal bed from the central Appalachian Basin [Ph.D. thesis]: Morgantown, West Virginia University, 495 p.

Eden, R. A., Elliot, R. W., Elliott, R. E., and Young, B. R., 1963, Tonstein bands in the coalfields of the East Midlands: Geological Magazine, v. 100, p. 47–58.

Fisher, R. V., and Schmincke, H. U., 1984, Pyroclastic rocks: Berlin, Springer-Verlag, 472 p.

Folinsbee, R. E., Baadsgaard, H., and Lipson, J., 1961, Potassium-argon dates of Upper Cretaceous ash falls, Alberta, Canada: New York Academy of Science Annals, v. 91, p. 352–359.

Folk, R. L., 1980, Petrology of sedimentary rocks: Austin, Texas, Hemphill Publishing, 170 p.

Forsman, N. F., 1984, Durability and alteration of some Cretaceous and Paleocene pyroclastic glasses in North Dakota: Journal of Non-Crystalline Solids, v. 67, p. 449–461.

Francis, E. H., 1961, Thin beds of graded kaolinized tuff and tuffaceous siltstone in the Carboniferous of Fife: Bulletin of the Geological Survey of Great Britain, v. 17, p. 191–215.

Francis, E. H., 1985, Recent ash fall: A guide to tonstein distribution: Congres International de Stratigraphie et Geologie du Carbonifere, 10th, Madrid, 1983, Comptes Rendus, v. 4, p. 189–195.

Francis, E. H., Smart, J.G.O., and Raisbeck, D. E., 1968, Westphalian volcanism at the horizon of the Black Rake in Derbyshire and Nottinghamshire: Yorkshire Geological Society Proceedings, v. 36, p. 395–416.

Garcia-Ramos, J., Aramburu, A., and Brime, C., 1984, Kaolin tonstein of volcanic ash origin in the Lower Ordovician of the Cantabrian Mountains (NW Spain): Trabajos de Geologia, v. 14, p. 27–33.

Heiken, G., and Wohletz, K., 1985, Volcanic ash: Berkeley, University of California Press, 246 p.

Hess, J. C., and Lippolt, H. J., 1986, $^{40}Ar/^{39}Ar$ ages of tonstein and tuff sanidines: New calibration points for the improvement of the Upper Carboniferous time scale: Chemical Geology (Isotope Geology Section), v. 59, p. 143–154.

Hess, J. C., Lippolt, H. J., and Burger, K., 1988, New time-scale calibration points in the Upper Carboniferous from Kentucky, Donetz basin, Poland, and West Germany [abs.]: Nuclear Tracks and Radiation Measurements, v. 17, p. 435–436.

Hodder, A.P.W., Green, B. E., and Lowe, D. J., 1990, A two-stage model for the formation of clay minerals from tephra-derived volcanic glass: Clay Minerals, v. 25, p. 313–327.

Hoehne, K., 1949, Neue Tonsteinvorkommen im Flozverband des Ruhrkarbons [New tonstein occurrences in the assemblage of seams of the Ruhr Carboniferous]: Gluckauf, v. 85, p. 756–757.

Hoehne, K., 1953a, Vorkommen von Kristalltonstein und Quarzneubildungen in tertiaren (?) Steinkohlenflozen von Oaxaca in Mexiko [Occurrences of crystalline tonstein and neogenic quartz in the Tertiary (?) coal seams of Oaxaca, Mexico]: Chemie der Erde, v. 16, p. 202–210.

Hoehne, K., 1953b, Kaolinkristalle und Quarzneubildungen in indischen Steinkohlen [Kaolin crystals and neogenic quartz in Indian bituminous coals]: Chemie der Erde, v. 16, p. 211–222.

Hoehne, K., 1957, Tonsteine in Kohlenflozen der Oststaaten von Nordamerika und Ostaustralien [Tonsteins in coal seams of the eastern states of North America and East Australia]: Chemie der Erde, v. 19, p. 111–129.

Hoehne, K., 1959, Grundsatzliche Erkenntnisse uber die Tonsteinbildung in Kohlenflozen und neue Tonsteinvorkommen in Ost-USA, Westkanada und Nordmexiko [Fundamental perceptions concerning tonstein formation in coal seams and new tonstein occurrences in eastern USA, western Canada, and northern Mexico]: Geologie, v. 8, p. 280–302.

Hoehne, K., 1964, Zur Entstehung und stratigraphischen Verbreitung der Kaolin-Kohlentonsteine in den wichtigsten Kohlenrevieren der Erde [On the origin and stratigraphic distribution of the kaolin coal tonsteins in the most important coal districts of the world]: Fortschritte in der Geologie von Rheinland und Westfalen, v. 12, p. 487–516.

Huff, W. D., Bergstrom, S. M., and Kolata, D. R., 1992, Gigantic Ordovician volcanic ash fall in North America and Europe: Biological, tectonomagmatic, and event-stratigraphic significance: Geology, v. 20, p. 875–878.

Izett, G. A., 1981, Volcanic ash beds: Recorders of upper Cenozoic silicic pyroclastic volcanism in the western United States: Journal of Geophysical Research, v. 86, p. 10200–10222.

Jeans, C. V., Merriman, R. J., and Mitchell, J. G., 1977, Origin of Middle Jurassic and Lower Cretaceous fuller's earth in England: Clay Minerals, v. 12, p. 11–44.

Keller, W. D., 1970, Environmental aspects of clay minerals: Journal of Sedimentary Petrology, v. 40, p. 788–813.

Kendrick, D. T., 1985, Vertical distribution of selected trace elements within the Fruitland Number Eight coal seam near Farmington, New Mexico [M.S. thesis]: Socorro, New Mexico Institute of Mining and Technology, 181 p.

Kennett, J. P., 1981, Marine tephrochronology, in Emiliani, C., ed., The oceanic lithosphere: The sea: New York, Wiley-Interscience, p. 1373–1436.

Kisch, H. J., 1966, Chlorite-illite tonstein in high-rank coals from Queensland, Australia: Notes on regional epigenetic grade and coal rank: American Journal of Science, v. 264, p 386–397.

Kolata, D. R., Frost, J. K., and Huff, W. D., 1986, K-bentonites of the Ordovician Decorah Subgroup, Upper Mississippi Valley: Correlation by chemical fingerprinting: Illinois State Geological Survey Circular 537, 30 p.

Kowallis, B. J., and Christiansen, E. J., 1989, Applications of zircon morphology: Correlation of pyroclastic rocks and petrogenic inferences: Geological Society of America Abstracts with Programs, v. 21, no. 6, p. A-244.

Kowallis, B. J., Christiansen, E. J., and Deino, A., 1989, Multicharacteristic correlation of Upper Cretaceous volcanic ash beds from southwestern Utah to central Colorado: Utah Geological and Mineral Survey Miscellaneous Publication 89, 22 p.

La Iglesia, A., and Van Oosterwyck-Gastuche, M. C., 1978, Kaolinite synthesis. I. Crystallization conditions at low temperatures and calculation of thermodynamic equilibria. Application to laboratory and field observations: Clays and Clay Minerals, v. 26, p. 397–408.

Lamb, H. H., 1970, Volcanic dust in the atmosphere; with a chronology and assessment of its meteorological significance: Royal Society of London Philosophical Transactions, v. 166, p. 425–533.

Lambrecht, L., and Scheere, J., 1965, Un tonstein d'age tertiaire dans le bassin charbonnier de Cali (Columbie, Amerique du Sud): Paris, Académie des Sciences Comptes Rendus, v. 260, p. 5310–5312.

Langway, C. C., Jr., Clausen, H. B., and Hammer, C. U., 1988, An interhemispheric volcanic time-marker in ice cores from Greenland and Antarctica: Annals of Glaciology, v. 10, p. 102–108.

Lerbekmo, J. F., Westgate, J. A., Smith, D.G.W., and Denton, G. H., 1975, New data on the character and history of the White River volcanic eruption, Alaska, in Creswell, M. M., ed., Quaternary studies: Wellington, Royal Society of New Zealand, p. 203–209.

Lippolt, H. J., and Hess, J. C., 1985, $^{40}Ar/^{39}Ar$ dating of sanidines from Upper Carboniferous tonsteins: Congres International de Stratigraphie et Geologie du Carbonifere, 10th, Madrid, 1983, Comptes Rendus, v. 4, p. 175–181.

Loughnan, F. C., 1971, Kaolinitic claystones associated with the Wongawilli Seam in the southern part of the Sydney Basin: Geological Society of Australia Journal, v. 18, p. 293–302.

Loughnan, F. C., 1978, Flint clays, tonsteins and the kaolinite clayrock facies: Clay Minerals, v. 13, p. 387–400.

Love, J. D., Duncan, D. C., Bergquist, H. R., and Hose, R. K., 1948, Stratigraphic sections of Jurassic and Cretaceous rocks in the Jackson Hole area, northwestern Wyoming: Wyoming Geological Survey Bulletin 40, 48 p.

Marvin, R. F., Bohor, B. F., and Mehnert, H. H., 1986, Tonsteins from New Mexico—Touchstones for dating coal beds: Isochron/West, v. 45, p. 17–18.

Masek, J., 1963, Produkte des oberkarbonischen Vulkanismus im Mittelboehmischen Kohlenbecken und das Entstehungsproblem der sog Tonsteine [Products of Upper Carboniferous volcanism in the central Bohemian coal basin and the problem of the origin of the so-called tonsteins]: Stuttgart, Neues Jahrbuch für Geologie und Palaontologie, Monatshefte, Stuttgart, p. 369–381.

McCabe, P. J., 1984, Depositional environments of coal and coal-bearing strata, in Rahmani, R. A., and Flores, R. M., eds., Sedimentology of coal and coal-bearing sequences (International Association of Sedimentologists Special Publication 7): Oxford, Blackwell Scientific Publications, p. 13–42.

Meriaux, E., 1972, Les tonsteins de la veine de charbon No. 10 (Balmer) a Sparwood Ridge dans le bassin de Fernie (Colombie Britannique): Geological Survey of Canada Report of Activities, Paper 72-1, part B, p. 11–22.

Moore, J. G., and Peck, D. L., 1962, Accretionary lapilli in volcanic rocks of the western continental United States: Journal of Geology, v. 70, p. 182–193.

Newell, R. E., and Walker, G.P.I., 1981, Volcanism and climate: Journal of Volcanology and Geothermal Research, v. 11, p. 1–92.

Norrish, K., 1968, Some phosphate minerals in soils (International Congress on Soil Science, 9th, Transactions, Volume 2): New York, Elsevier, p. 713–723.

Olejnik, S., Aylmore, A. G., Posner, A. M., and Quirk, J. P., 1968, Infrared spectra of kaolin mineral-dimethyl sulfoxide complexes: Journal of Physical Chemistry, v. 72, p. 241–249.

Outerbridge, W. F., Triplehorn, D. M., and Lyons, P. C., 1990, The Princess #6 Middle Pennsylvanian volcanic ash fall (tonstein), Kentucky and West Virginia, central Appalachian Basin: Southeastern Geology, v. 31, p. 63–78.

Patterson, S. H., and Hosterman, J. W., 1958, Geology of the clay deposits in the Olive Hill district, Kentucky, in Swineford, A., ed., Clays and Clay Minerals, Proceedings of the 7th National Conference: New York, Pergamon Press, p. 178–194.

Pevear, D. R., Williams, V. E., and Mustoe, G. E., 1980, Kaolinite, smectite, and K-rectorite in bentonites: Relation to coal rank at Tulameen, British Columbia: Clays and Clay Minerals, v. 28, p. 241–254.

Pollastro, R. M., 1981, Authigenic kaolinite and associated pyrite in chalk of the Cretaceous Niobrara Formation, eastern Colorado: Journal of Sedimentary Petrology, v. 51, p. 553–562.

Pollastro, R. M., 1983, The formation of illite at the expense of illite/smectite: Mineralogical and morphological support for a hypothesis [abs], in Program and Abstracts, The Clay Minerals Society, Annual Clay Minerals Conference, 22nd, Buffalo, New York: Buffalo, State University of New York, Department of Geological Sciences, p. 82.

Pollastro, R. M., and Bohor, B. F., 1993, Origin and genesis of the Cretaceous/Tertiary boundary claystone, Western Interior of North America: Clays and Clay Minerals, v. 41 (in press).

Pollastro, R. M., and Martinez, C. J., 1985, Whole-rock, insoluble residue, and clay mineralogies of marl, chalk, and bentonite, Smoky Hill Shale Member, Niobrara Formation near Pueblo, Colorado—Depositional and diagenetic implications, in Pratt, L. M., Kauffman, E. G., and Zelt, F. B., eds., Fine-grained deposits and biofacies of the Western Interior seaway;

evidence of cyclic sedimentary processes (Society of Economic Paleontologists and Mineralogists, Annual Midyear Meeting, 2nd, Guidebook): Tulsa, Oklahoma, Society of Economic Paleontologists and Mineralogists, p. 215–222.

Pollastro, R. M., and Pillmore, C. L., 1987, Mineralogy and petrology of the Cretaceous/Tertiary boundary clay bed and adjacent clay-rich rocks, Raton Basin, New Mexico and Colorado: Journal of Sedimentary Petrology, v. 57, p. 456–466.

Pollastro, R. M., Pillmore, C. L., Tschudy, R. H., Orth, C. J., and Gilmore, J. S., 1983, Clay petrology of the conformable Cretaceous/Tertiary boundary interval, Raton Basin, New Mexico and Colorado [abs.], in Program with Abstracts, Clay Minerals Society, Annual Meeting, 20th, Buffalo, New York: Buffalo, State University of New York, Department of Geological Science, p. 83–84.

Porter, S. C., 1981, Recent glacier variations and volcanic eruptions: Nature, v. 291, p. 139–142.

Prado, J. G., 1964, Consideration sur quelques particularites genetiques des premiers tonstein decouverts dans le bassin houiller des Asturies (Espagne): Congres International de Stratigraphie et de Geologie du Carbonifere, Paris and Grenoble, 1963, Comptes Rendus, Part 2, p. 693–704.

Price, N. B., and Duff, P., 1969, Mineralogy and chemistry of tonsteins from Carboniferous sequences in Great Britain: Sedimentology, v. 13, p. 45–69.

Puffer, J. H., Russell, E.W.B., and Rampino, R. R., 1980, Distribution and origin of magnetite spherules in air, waters, and sediments of the greater New York City area and the North Atlantic Ocean: Journal of Sedimentary Petrology, v. 50, p. 247–256.

Pupin, J., 1980, Zircon and granite petrology: Contributions to Mineralogy and Petrology, v. 73, p. 207–220.

Reinink-Smith, L. M., 1982, The mineralogy, geochemistry and origin of bentonite partings in the Eocene Skookumchuck Formation, Centralia Mine, southwestern Washington [M.S. thesis]: Bellingham, Western Washington Unviersity, 119 p.

Reinink-Smith, L. M., 1990a, Mineral assemblages of volcanic and detrital partings in Tertiary coal beds, Kenai Peninsula, Alaska: Clays and Clay Minerals, v. 38, p. 97–108.

Reinink-Smith, L. M., 1990b, Relative frequency of Neogene volcanic events as recorded in coal partings from the Kenai lowland, Alaska: A comparison with deep-sea core data: Geological Society of America Bulletin, v. 102, p. 830–840.

Richardson, G., and Francis, E. H., 1971, Fragmental clayrock (FCR) in coal-bearing sequences in Scotland and Northeast England: Yorkshire Geological Society Proceedings, v. 38, p. 229–260.

Rogers, G. S., 1914, The occurrence and genesis of a persistent parting in a coal bed of the Lance Formation: American Journal of Science, v. 87, p. 299–304.

Ross, C. S., 1928, Altered Paleozoic volcanic minerals and their recognition: American Association of Petroleum Geologists Bulletin, v. 12, p. 143–164.

Ross, C. S., and Shannon, E. V., 1926, The minerals of bentonite and related clays and their physical properties: Journal of the American Ceramic Society, v. 9, p. 77–96.

Ruddiman, W. F., and Glover, L. K., 1972, Vertical mixing of ice-rafted volcanic ash in North Atlantic sediments: Geological Society of America Bulletin, v. 83, p. 2817–2836.

Ruppert, L. F. Cecil, C. B., and Stanton, R. W., 1984, Authigenic quartz in the Upper Freeport coal bed, west-central Pennsylvania: Journal of Sedimentary Petrology, v. 55, p. 334–339.

Ryer, T. A., Phillips, R. E., Bohor, B. F., and Pollastro, R. M., 1980, Use of altered volcanic ash falls in stratigraphic studies of coal-bearing sequences: An example from the Upper Cretaceous Ferron Sandstone Member of the Mancos Shale in central Utah: Geological Society of America Bulletin, v. 91 part 1, p. 579–586.

Samson, S. D., Kyle, P. R., and Alexander, E. C., 1988, Correlation of North American Ordovician bentonites by using apatite chemistry: Geology, v. 6, p. 444–447.

Samson, S. D., Patchett, P. J., Roddick, C. J., and Parrish, R. R., 1989, Origin and tectonic setting of Ordovician bentonites in North America: Isotopic and age constraints: Geological Society of America Bulletin, v. 101, p. 1175–1181.

Sanchez, J. D., Bradbury, J. P., Bohor, B. F., and Coates, D. A., 1987, Diatoms and tonsteins as paleoenvironmental and paleodepositional indicators in a Miocene coal bed, Costa Rica: Palaios, v. 2, p. 158–164.

Scheere, J., 1956, Nouvelle Contribution a le'etude des Tonstein du Terrain houllier Belge [New contribution to the study of the tonsteins from the Belgian coal measures]: Bruxelles, Association pour l'Etude de Paleontologie et Stratigraphie, v. 26, p. 1–52.

Schmitz-Dumont, W., 1894, Die Saarbrucker Tonsteine: Wilhelmshaven, Tonindustrie-Zeitung, v. 18, p. 714.

Schultz, L. G., 1963, Nonmontmorillonitic composition of some bentonite beds, in Clays and Clay Minerals, Proceedings of the 11th National Conference: New York, Pergamon Press, p. 169–177.

Schumacher, R., 1988, Aschenaggregate in vulkaniklastischen Transportsystemen [Ph.D. thesis]: Bochum, Germany, Ruhr-Universität, 147 p.

Seiders, V. M., 1965, Volcanic origin of flint clay in the Fire Clay coal bed, Breathitt Formation, Eastern Kentucky, in Geological Survey Research, Chapter D; U.S. Geological Survey Professional Paper 525-D, p. D52–D54.

Senkayi, A. L., Dixon, J. B., Hossner, L. R., Abder-Ruhman, M., and Fanning, D. S., 1984, Mineralogy and genetic relationships of tonstein, bentonite, and lignitic strata in the Eocene Yegua Formation of east-central Texas: Clays and Clay Minerals, v. 32, p. 259–271.

Senkayi, A. L., Ming, D. W., Dixon, J. B., and Hossner, L. R., 1987, Kaolinite, opal-CT, and clinoptilolite in altered tuffs interbedded with lignite in the Jackson Group, Texas: Clays and Clay Minerals, v. 35, p. 281–290.

Skocek, V., 1973, Contribution to the problem of tonstein origin: Casopis pro Mineralogii a Geologii, v. 18, p. 233–242.

Slaughter, M., and Earley, J. W., 1965, Mineralogy and geological significance of the Mowry bentonites, Wyoming: Geological Society of America Special Paper 83, 116 p.

Spears, D. A., 1966, A Westphalian tonstein from south Staffordshire: Yorkshire Geological Society Proceedings, v. 35, p. 523–548.

Spears, D. A., 1971, The mineralogy of the Stafford tonstein: Yorkshire Geological Society Proceedings, v. 38, p. 497–516.

Spears, D. A., 1982, The recognition of volcanic clays and the significance of heavy minerals: Clay Minerals, v. 17, p. 373–375.

Spears, D. A., and Duff, P. M., 1984, Kaolinite and mixed-layer illite-smectite in Lower Cretaceous bentonites from the Peace River coal field, British Columbia: Canadian Journal of Earth Sciences, v. 21, p. 465–476.

Spears, D. A., and Kanaris-Sotiriou, R., 1975, Titanium in some Carboniferous sediments from Great Britain: Geochimica et Cosmochimica Acta, v. 40, p. 345–351.

Spears, D. A., Duff, P. M., and Caine, P. M., 1988, The West Waterberg tonstein, South Africa: International Journal of Coal Geology, v. 9, p. 221–233.

Spears, D. A., and Kanaris-Sotiriou, R., 1979, A geochemical and mineralogical investigation of some British and other European tonsteins: Sedimentology, v. 26, p. 407–425.

Spears, D. A., and Rice, D. M., 1973, An Upper Carboniferous tonstein of volcanic origin: Sedimentology, v. 20, p. 281–294.

Srodon, J., 1976, Mixed-layer smectite/illites in the bentonites and tonsteins of the Upper Silesian coal basin: Polska Academia Nauk Oddzial Krakowie; Komisja Nauk Mineralogicnych, Prace Mineral, no. 49, 84 p.

Stach, E., 1950, Vulkanische Aschenregen uber dem Steinkohlenmoor [Volcanic ashfalls upon the bituminous coal swamp]: Gluckauf, v. 86, p. 1–50.

Stach, E., Taylor, G. H., Mackausky, M. T., Chandra, D., Teichmuller, M., and Teichmuller, R., 1982, Stach's textbook of coal petrology: Berlin,

Stadler, G., and Werner, H., 1962, Ein Phosphat-Mineral der Crandallit-Gruppe in den Kaolin-Kohlenstoneinen des Ruhrkarbons: Fortschritte in der Geologie von Rheinland und Westfalen, v. 3, p. 619–622.

Stevens, S. S., 1979, Petrogenesis of a tonstein in the Appalachian bituminous basin [M.S. thesis]: Richmond, Eastern Kentucky University, 82 p.

Stöffler, D., 1963, Neure Erkenntnisse in der Tonsteinfrage auf Grund sedimentpetrographisher und geochemischer Untersuchungen im Floz Wahlschied der Grube Ensdorf (Saar): Beitrage zur Mineralogie und Petrographie, v. 9, p. 285–312.

Strauss, P. G., 1971, Kaolin-rich rocks in the East Midlands coalfields of England: Congres International de Stratigraphie et Geologie du Carbonifere, 6th, Sheffield, Comptes Rendus, 1967, v. 4, p. 1519–1532.

Stutzer, O., 1931, Vulkanische Aschen als Leitlagen in Kohlenflozen: Zeitschrift fur Praktische Geologie, v. 10, p. 145–148.

Tan, K. H., ed., 1984, Andolsols (Benchmark Papers in Soil Science, Volume 4): New York, Van Nostrand Reinhold, 418 p.

Tereshenko, S. P. and Chernovyantz, M. G., 1979, The first discovery of tonstein in coal measures of the L'vov-Volensky Basin: Geologischniy Zhurnal (Kiev), v. 39, no. 1, p. 94–98.

Thorez, J., and Pirlet, H., 1979, Petrology of K-bentonite beds in the carbonate series of the Visean and Tournaisian stages of Belgium, in Mortland, M. M., and Farmer, V. C., eds., Proceedings, International Clay Conference, 1978 (Developments in Sedimentology, Volume 27): New York, Elsevier Science Publishing Co., p. 323–332.

Triplehorn, D. M., 1976a, Contributions to clay mineralogy and petrology, Cook Inlet basin, Alaska: Alaska Division of Geological and Geophysical Surveys Open-File Report 102, 19 p.

Triplehorn, D. M., 1976b, Volcanic ash partings in coals: Characteristics and stratigraphic significance, in Fritsch, A. E., TerBest, H., Jr., and Wornardt, W. W., eds., The Neogene Symposium; papers presented at the Pacific Section: San Francisco, California, American Association of Petroleum Geologists, p. 9–12.

Triplehorn, D. M., and Bohor, B. F., 1981, Altered volcanic ash partings in the C coal, Ferron Sandstone Member of the Mancos Shale, Emery County, Utah: U.S. Geological Survey Open-File Report 81-775, 43 p.

Triplehorn, D. M., and Bohor, B. F., 1983, Goyazite in kaolinitic altered tuff beds of Cretaceous age near Denver, Colorado: Clays and Clay Minerals, v. 31, p. 299–304.

Triplehorn, D. M., and Bohor, B. F., 1986, Volcanic ash layers in coal: Origin, distribution, composition and significance, in Vorres, K. S., ed., Mineral matter and ash in coal (Symposium Series No. 301): Washington, D.C., American Chemical Society, p. 90–98.

Triplehorn, D. M., and Finkleman, R. B., 1989, Replacement of glass shards by aluminum phosphates in a Middle Pennsylvanian tonstein from eastern Kentucky: Geological Society of America Abstracts with Programs, v. 21, no. 6, p. A52–A53.

Triplehorn, D. M., Turner, D. L., and Naeser, C. W., 1977, K-Ar and fission-track dating of ash partings in Tertiary coals from the Kenai Penninsula, Alaska: A radiometric age for the Homerian-Clamgulchian Stage boundary: Geological Society of America Bulletin, v. 88, p., 1156–1160.

Triplehorn, D. M., Turner, D. L., and Naeser, C. W., 1984, Radiometric age of the Chickaloon formation, south-central Alaska: Location of the Paleocene-Eocene boundary: Geological Society of America Bulletin, v. 95, p. 740–742.

Triplehorn, D. M., Outerbridge, W. F., and Lyons, P. C., 1989, Six new altered volcanic ash beds (tonsteins) in the Middle Pennsylvanian of the Appalachian basin, Virginia, West Virginia, Kentucky, and Ohio: Geological Society of America Abstracts with Programs, v. 21, no. 6, p. A134.

Triplehorn, D. M., Stanton, R. W., Ruppert, L. F., and Crowley, S. S., 1991, Volcanic ash dispersed in the Wyodak-Anderson coal bed, Powder River basin, Wyoming: Organic Geochemistry, v. 17, p. 567–575.

Triplehorn, D. M., Bohor, B. F., and Betterton, W. J., 1993, Chemical disaggregation of kaolinitic tonsteins and flint clays: Clays and Clay Minerals, v. 41 (in press).

Turner, D. L., Triplehorn, D. M., Naeser, C. W., and Wolfe, J. A., 1980, Radiometric dating of ash partings in Alaskan coal beds and upper Tertiary paleobotanical stages: Geology, v. 8, p. 92–96.

Turner, D. L., Triplehorn, D. M., Frizzell, V. A., and Naeser, C. W., 1983, Radiometric dating of ash partings in coals of the Eocene Puget Group, Washington: Implication for paleobotanical studies: Geology, v. 11, p. 527–531.

Waage, K. M., 1961, Stratigraphy and refractory clayrocks of the Dakota Group along the northern Front Range, Colorado: U.S. Geological Survey Bulletin 1102, 154 p.

Warwick, P. D., and Stanton, R. W., 1988, Petrographic characteristics of the Wyodak-Anderson coal bed (Paleocene), Powder River Basin, Wyoming, U.S.A: Organic Geochemistry, v. 12, p. 389–399.

Weaver, C. E., 1963, Interpretative value of heavy minerals from bentonites: Journal of Sedimentary Petrology, v. 33, p. 343–349.

Weiss, A., Thielepape, W., Ritter, W., Schafer, H., and Goring, G., 1963, Zur Kenntnis von Hydrazin-Kaolinit: Zeitschrift für Anorganische und Allgemeine Chemie, v. 320, p. 183–204.

Williamson, I. A., 1961, Tonsteins: A possible aid to coal field correlation: Mining Magazine, v. 104, p. 9–14.

Williamson, I. A., 1970a, Tonsteins—Their nature, origins and uses: Mining Magazine, v. 122, no. 2, p. 119–225.

Williamson, I. A., 1970b, Tonsteins—Their nature, origins and uses: Mining Magazine, v. 122, no. 3, p. 208–211.

Wilson, A. A., Sergeant, G. A., Young, B. R., and Harrison, R. K., 1966, The Rowhurst tonstein, North Staffordshire, and the occurrence of crandallite: Yorkshire Geological Society Proceedings, v. 35, p. 421–427.

Zaritsky, P. V., 1967, Origin of tonsteins within Donbass coal seams: Report of the USSR Academy of Sciences, v. 177, no. 2, 422–425.

Zaritsky, P. V., 1971, Kaolinitic claystone partings in Donetz coal seams and their significance in correlation: Osadkonakopleniye i genezis uglay Karbona SSSR, Nauka Press (Moscow), p. 151–162.

Zaritsky, P. V., 1985, A review of the study of tonsteins in the Donetz basin: Congres International de Stratigraphie et Geologie du Carbonifere, 10th, Madrid, 1983, Comptes Rendus, v. 4, p. 235–241.

Zhou, Y., Ren, Y., and Bohor, B. F., 1982, Origin and distribution of tonsteins in Late Permian Coal seams of southwestern China: International Journal of Coal Geology, v. 2, p. 49–77.

Zielinski, R. A., 1982, The mobility of uranium and other elements during alteration of rhyolitic ash to montmorillonite: A case study in the Troublesome Formation, Colorado: Chemical Geology, v. 35, p. 185–204.

Zielinski, R. A., 1985, Element mobility during alteration of silicic ash to kaolinite—A study of tonstein: Sedimentology, v. 32, p. 567–579.

Manuscript Accepted by the Society January 6, 1993